Lecture Notes in Electrical Engineering

Volume 485

The book series *Lecture Notes in Electrical Engineering* (LNEE) publishes the latest developments in Electrical Engineering—quickly, informally and in high quality. While original research reported in proceedings and monographs has traditionally formed the core of LNEE, we also encourage authors to submit books devoted to supporting student education and professional training in the various fields and applications areas of electrical engineering. The series cover classical and emerging topics concerning:

- Communication Engineering, Information Theory and Networks
- Electronics Engineering and Microelectronics
- Signal, Image and Speech Processing
- Wireless and Mobile Communication
- Circuits and Systems
- Energy Systems, Power Electronics and Electrical Machines
- Electro-optical Engineering
- Instrumentation Engineering
- Avionics Engineering
- Control Systems
- Internet-of-Things and Cybersecurity
- Biomedical Devices, MEMS and NEMS

For general information about this book series, comments or suggestions, please contact leontina.dicecco@springer.com.

To submit a proposal or request further information, please contact the Publishing Editor in your country:

China

Jasmine Dou, Editor (jasmine.dou@springer.com)

India, Japan, Rest of Asia

Swati Meherishi, Editorial Director (Swati.Meherishi@springer.com)

Southeast Asia, Australia, New Zealand

Ramesh Nath Premnath, Editor (ramesh.premnath@springernature.com)

USA, Canada

Michael Luby, Senior Editor (michael.luby@springer.com)

All other Countries

Leontina Di Cecco, Senior Editor (leontina.dicecco@springer.com)

**** This series is indexed by EI Compendex and Scopus databases. ****

Alfonso Bachiller Soler · Ramón Cano González ·
Miguel Angel González Cagigal

Theorems for Electrical Circuits

Theory and Solved Problems

Springer

Alfonso Bachiller Soler
Department of Electrical Engineering
University of Seville
Sevilla, Spain

Ramón Cano González
Department of Electrical Engineering
University of Seville
Sevilla, Spain

Miguel Angel González Cagigal
Department of Electrical Engineering
University of Seville
Sevilla, Spain

ISSN 1876-1100 ISSN 1876-1119 (electronic)
Lecture Notes in Electrical Engineering
ISBN 978-3-031-74908-7 ISBN 978-3-031-74909-4 (eBook)
https://doi.org/10.1007/978-3-031-74909-4

Translation from the Spanish language edition: "Circuitos Eléctricos. Teoremas—Teoría y Problemas resueltos" by Alfonso Bachiller Soler et al., © Ediciones Díaz de Santos 2024. Published by Ediciones Díaz de Santos. All Rights Reserved.

This Springer imprint is published by the registered company Springer Nature Switzerland AG
The registered company address is: Gewerbestrasse 11, 6330 Cham, Switzerland

To our students

Foreword

Circuit Theory is one of the basic sciences of Engineering, so the study of its fundamental concepts is present in all undergraduate engineering studies under some of its multiple denominations (Circuit Theory, Electrical Engineering, Electrical Technology, etc.). In this sense, there is a wide variety of books that address the basic concepts of Circuit Theory, usually with a certain focus according to the students to whom the work is directed, which, in some way, distorts the knowledge of the general principles that govern any electrical circuit.

This work addresses a core part of Circuit Theory, such as its fundamental theorems: consequences of linearity in electrical circuits, equivalent circuits, maximum power transfer, as well as the applications of magnetic coupling in circuits. It is, therefore, a book of special interest as it is focused on the theoretical and practical study of the properties of electrical circuits, a fundamental basis for delving into other more practical aspects, such as transient and steady-state regimes, equations for circuit resolution, alternating and three-phase circuits, etc.

The authors of this work, Alfonso Bachiller Soler, Ramón Cano González, and Miguel Angel González Cagigal, are Ph.Ds. in engineering with extensive teaching experience in the Department of Electrical Engineering at the University of Seville. Their deep knowledge of Circuit Theory is not only due to their accredited teaching experience in various undergraduate engineering programs but also to their relevant contributions to research and technology transfer to the industry, contributions that demonstrate their mastery of Circuit Theory as a basic science of Engineering, especially Electrical Engineering. This experience supports the elaboration of this textbook, which undoubtedly constitutes a work with the required rigor and of great educational value for engineering students who must tackle the study of the basic concepts of electrical circuits.

July 2024

José Luis Martínez Ramos
Full Professor
University of Seville
Sevilla, Spain

Preface

This text is intended for an introductory course on electric circuits, with a focus on applying theorems to circuit analysis, typically covered in the second course of an electrical engineering degree or similar programs. Specifically, the superposition, Thévenin, Norton, and maximum power transfer theorems are addressed.

The book includes a theoretical section followed by a substantial number of solved problems. Each exercise is resolved in detail, supported by more than 300 figures.

Sevilla, Spain
July 2024

Alfonso Bachiller Soler
Ramón Cano González
Miguel Angel González Cagigal

Contents

Chapter 1
Theorems

Abstract When solving an electrical circuit, there are a series of theorems that allow simplifying the calculation and help to have a better understanding of its operation. Such is the case of the superposition theorem, Thévenin's theorem, Norton's theorem, or maximum power transfer theorem, which are the ones that will be discussed in this book. First, we will recall the linearity property applied to electrical circuits, and then we will describe a series of theorems applicable to linear circuits.

1.1 Linear Circuits

Linearity is a property of an element that describes a linear relationship between cause and effect. This characteristic is a combination of the properties of homogeneity (proportionality) and additivity. The property of proportionality states that if the excitation (independent sources) is multiplied by a constant, the response (any voltage or current of a particular element) is also multiplied by the same constant. The additive property states that the response to a sum of excitation is equal to the sum of the responses to each excitation separately.

Consider a linear circuit where x denotes the value of the excitation, and y denotes the voltage or current of a particular element in that circuit. The relationship between cause and effect (between excitation and response) can be expressed mathematically as:

$$y = f(x) \tag{1.1}$$

Since it is a linear circuit, then $f(x)$ must be a linear function, and therefore, it will have the properties of proportionality and additivity:

1. $f(kx) = kf(x)$
2. $f(x_1 + x_2) = f(x_1) + f(x_2)$

Both properties can be combined as follows:

$$f(k_1x_1 + k_2x_2) = k_1 f(x_1) + k_2 f(x_2) \tag{1.2}$$

A. Bachiller Soler et al., *Theorems for Electrical Circuits*, Lecture Notes in Electrical Engineering 485, https://doi.org/10.1007/978-3-031-74909-4_1

In the case, for example, of a resistor (a linear element), the voltage-current relationship is $u = Ri$, so $f(i) = Ri$. If

$$i = k_1 i_1 + k_2 i_2 \tag{1.3}$$

then it can be verified that it satisfies the properties of proportionality and additivity:

$$f(k_1 i_1 + k_2 i_2) = R(k_1 i_1 + k_2 i_2) = k_1 R i_1 + k_2 R i_2 = k_1 f(i_1) + k_2 f(i_2) \tag{1.4}$$

However, the power dissipated in a resistor is not a linear function of either current or voltage. If $p = Ri^2$ represents the power in a resistor, expressed as a function of current, then $f(i) = Ri^2$. If

$$i = k_1 i_1 + k_2 i_2 \tag{1.5}$$

then:

$$f(k_1 i_1 + k_2 i_2) = R(k_1 i_1 + k_2 i_2)^2 \neq k_1 R i_1^2 + k_2 R i_2^2 \tag{1.6}$$

The same conclusion is reached if the power is expressed as a function of the resistor's voltage.

In summary, a circuit is linear if it consists solely of linear elements: resistors, inductors, capacitors, dependent sources, independent sources, etc. And therefore, it satisfies the properties of proportionality and additivity (which will be further elaborated in the superposition theorem).

1.2 Superposition and Proportionality

In a linear circuit composed of passive elements, dependent sources, and independent sources [1], the response (voltage and/or current) due to the simultaneous action of all independent sources is equal to the sum (superposition) of the partial responses obtained by separately applying each of these independent sources, while the rest of the independent sources are nullified (Fig. 1.1). It is worth noting that an independent voltage source is nullified by short-circuiting its terminals, while an independent current source is nullified by leaving its terminals open-circuited. Dependent sources are never nullified.

Figure 1.2 illustrates the application of the principles of proportionality and superposition to a particular generic circuit. Thus, the voltage $u(t)$ and the current $i(t)$ are obtained as follows:

$$u(t) = \alpha u_1(t) + \beta u_2(t) \ ; \quad i(t) = \alpha i_1(t) + \beta i_2(t) \tag{1.7}$$

Superposition is applied both in the analysis and design of circuits [3]. When analyzing a complex circuit with multiple independent voltage and current sources, there will typically be fewer equations, and they will be simpler to solve when the

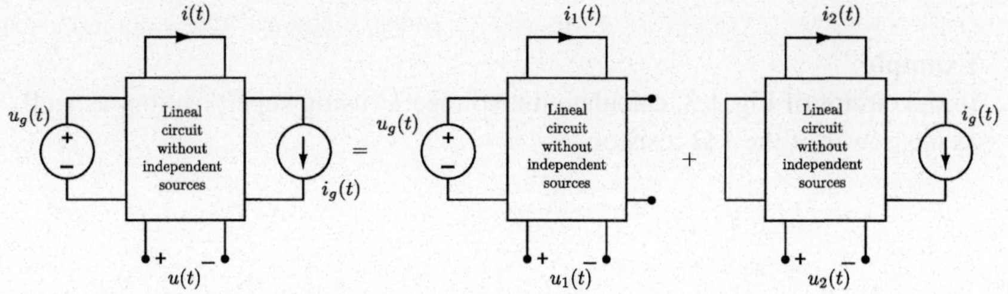

Fig. 1.1 Application of superposition [2]

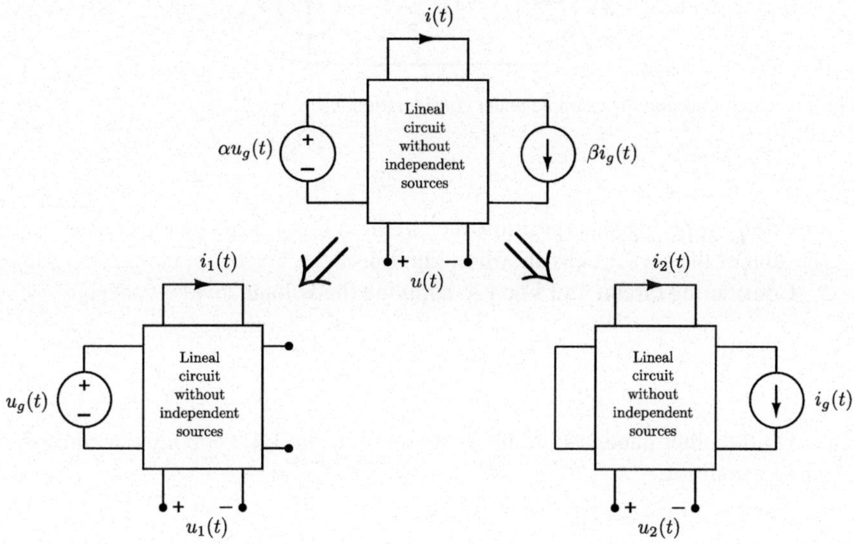

Fig. 1.2 Application of proportionality and superposition

effects of the different independent sources are considered separately. Therefore, the application of superposition can simplify the analysis of circuits. However, there are circuits where the use of superposition is necessary, such as circuits excited by AC sources of different frequencies or circuits excited by both AC and DC sources.

In the case of design, superposition is applied to synthesize a desired response from a circuit that cannot be achieved using a single source. This technique allows working with several simple designs instead of a complex one.

An important aspect to consider is that superposition is based on linearity. As mentioned earlier, the total power will not be correctly obtained by summing the powers due to each source acting separately. However, superposition can be indirectly used to calculate power since both voltage and current can be calculated using superposition. The application of superposition will also not lead to correct responses for circuits containing nonlinear elements, such as certain electronic elements like diodes.

Example
In the circuit of Fig. 1.3, calculate the voltage U using superposition, as well as the power of the 1 Ω resistor.

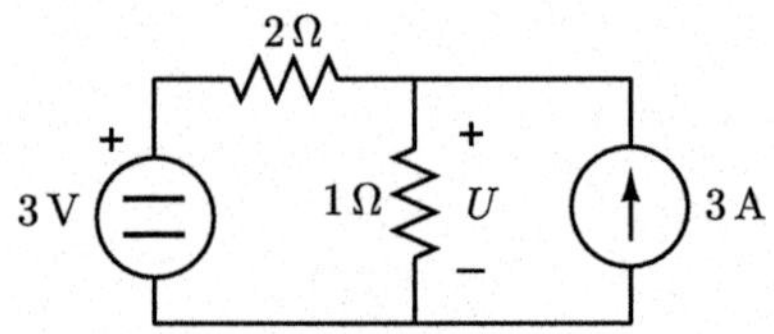

Fig. 1.3 Circuit used in the example under consideration

Solution Applying superposition, the circuit in Fig. 1.3 can be represented as the sum of the two circuits shown in Fig. 1.4.

Considering circuit 1 of Fig. 1.4, applying the voltage divider concept:

$$U_1 = 3 \cdot \frac{1}{2+1} = 1\,\mathrm{V}$$

On the other hand, according to circuit 2 of Fig. 1.4, applying the current divider concept:

$$I_2 = 3 \cdot \frac{1}{1+1/2} = 2\,\mathrm{A} \Rightarrow U_2 = 1 \cdot I_2 = 1 \cdot 2 = 2\,\mathrm{V}$$

Consequently, applying superposition, the voltage U is as follows:

$$U = U_1 + U_2 = 1 + 2 = 3\,\mathrm{V}$$

The power of the 1 Ω resistor is as follows:

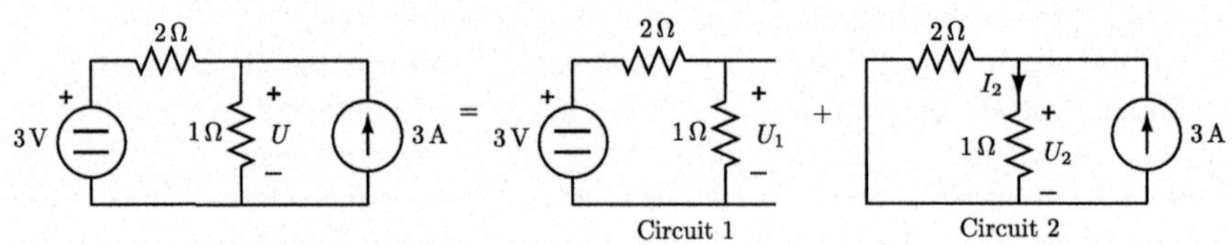

Fig. 1.4 Application of superposition theorem

$$p = \frac{U^2}{1} = \frac{3^2}{1} = 9\,\text{W}$$

It can be verified that, due to the nonlinearity of power, if its value in the 1 Ω resistor is calculated by summing the power absorbed in circuit 1 and the power absorbed in circuit 2, the erroneous result is as follows:

$$p' = p_1 + p_2 = \frac{U_1^2}{1} + \frac{U_2^2}{1} = \frac{1^2}{1} + \frac{2^2}{1} = 5\,\text{W}$$

1.3 Equivalent Circuits

Sometimes it is not necessary to perform a complete analysis of a circuit, but only to understand the behavior of one or more of its elements. In this case, the use of equivalent circuits is very useful. An equivalent circuit is one whose voltage-current characteristic is identical to that of the original circuit.

Figure 1.5 shows an original linear circuit with a certain voltage-current characteristic at its terminals $f_o[u(t), i(t)]$. This original linear circuit is to be replaced by an equivalent linear circuit [2]. In this sense, it is necessary for the voltage-current characteristic of the equivalent linear circuit ($f_e[u(t), i(t)]$) to match that of the original linear circuit, regardless of the external circuit:

$$f_o[u(t), i(t)] = f_e[u(t), i(t)] \tag{1.8}$$

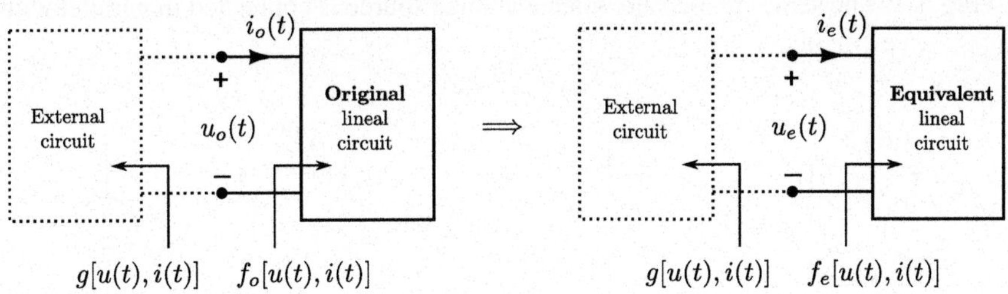

Fig. 1.5 Equivalent circuits

By applying the concept of equivalent circuit, the relationships between the parameters of a real voltage source and a real current source can be obtained so that they are equivalent to each other. Figure 1.6 shows a real DC voltage source and a real DC current source.

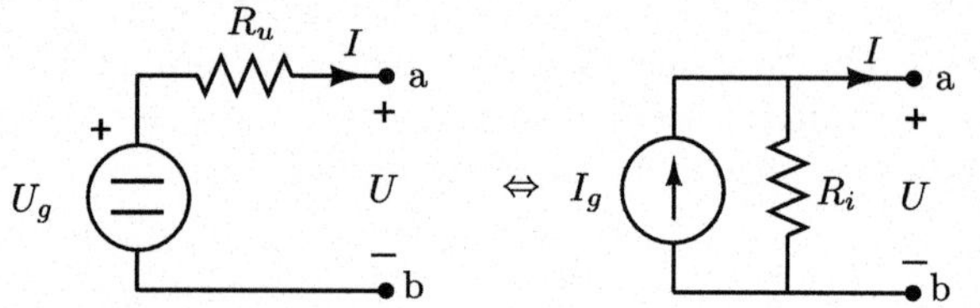

Fig. 1.6 Equivalence of real sources

The voltage-current relationship from terminals a and b of the real voltage source is as follows:

$$U = U_g - R_u \cdot I \ \Rightarrow \ U + R_u \cdot I - U_g = 0 \tag{1.9}$$

On the other hand, the voltage-current relationship from terminals a and b of the real current source is as follows:

$$I = I_g - \frac{U}{R_i} \ \Rightarrow \ U + R_i \cdot I - R_i \cdot I_g = 0 \tag{1.10}$$

For the real sources in Fig. 1.6 to be equivalent, both relationships must be equal. Therefore:

$$R_u = R_i = R \ ; \ U_g = R \cdot I_g \tag{1.11}$$

There are circuit analysis problems that are simplified by source transformations, i.e., replacing a real voltage source with its equivalent real current source or viceversa.

In the case of a voltage source connected in parallel with a current source, the current source has no effect on the equivalent circuit from terminals a and b as shown in Fig. 1.7. The same can be said when a voltage source is connected in parallel with a passive element.

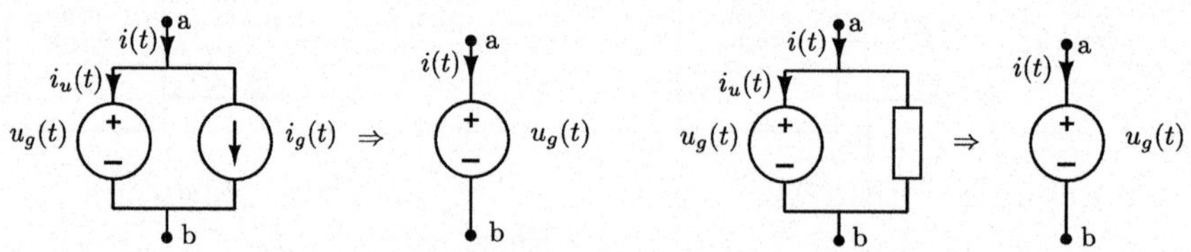

Fig. 1.7 Equivalent circuits containing a voltage source in parallel with a passive element or with an active element

Similarly, in the case of a current source connected in series with a voltage source, the voltage source has no effect on the equivalent circuit from terminals a and b, as shown in Fig. 1.8. The same applies when a current source is connected in series with a passive element.

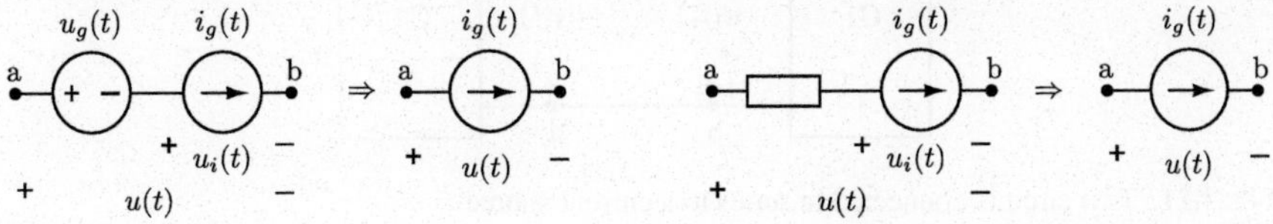

Fig. 1.8 Equivalent circuits containing a current source in series with a passive element or with an active element

It is important to note that, according to Figs. 1.7 and 1.8, neither the equivalent voltage source nor the equivalent current source are the original sources.

Next, we will analyze the case where two circuits (C1 and C2) are connected in parallel to a voltage source as shown in Fig. 1.9. Taking into account the equivalents shown in Fig. 1.7, then the circuit in Fig. 1.9 can be analyzed by separately solving circuit C1 and circuit C2 together with the voltage source, as shown in Fig. 1.10. Note that the current flowing through the voltage source is calculated as the sum of $i_1(t)$ and $i_2(t)$.

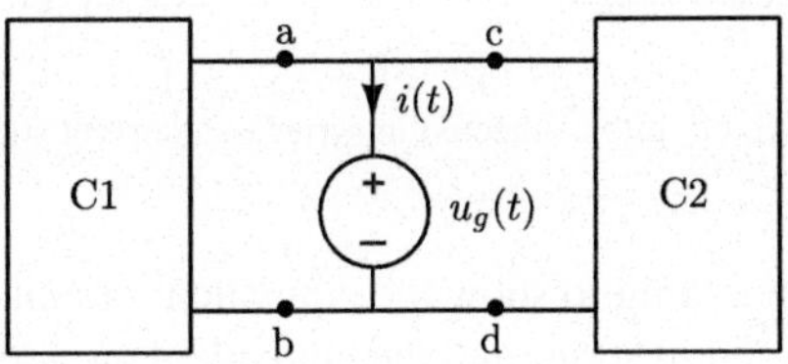

Fig. 1.9 Two circuits connected in parallel to a voltage source

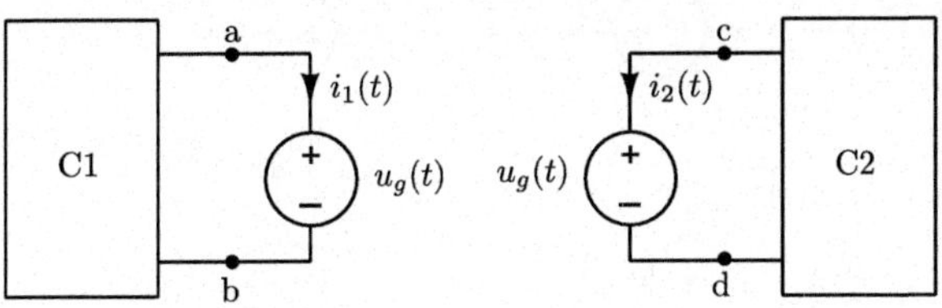

Fig. 1.10 Equivalent of two circuits connected in parallel to a voltage source

Similarly, in Fig. 1.11, two circuits (C1 and C2) are shown connected in series with a current source.

Taking into account the equivalents shown in Fig. 1.8, then the circuit in Fig. 1.11 can be analyzed by separately solving circuit C1 and circuit C2 together with the

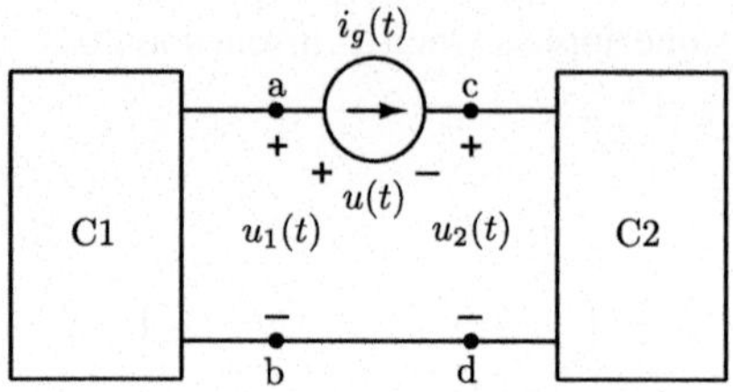

Fig. 1.11 Two circuits connected in series to a current source

current source, as shown in Fig. 1.12. Note that the voltage of the current source is calculated from the voltages $u_1(t)$ and $u_2(t)$:

$$u(t) = u_1(t) - u_2(t) \tag{1.12}$$

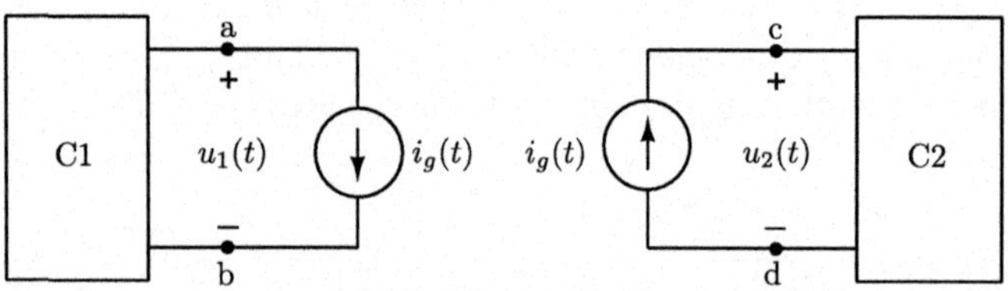

Fig. 1.12 Equivalent of two circuits connected in series to a current source

Finally, it is also interesting to show the equivalent circuit in the case of multiple voltage sources connected in series and the equivalent circuit in the case of multiple current sources connected in parallel, as shown in Fig. 1.13.

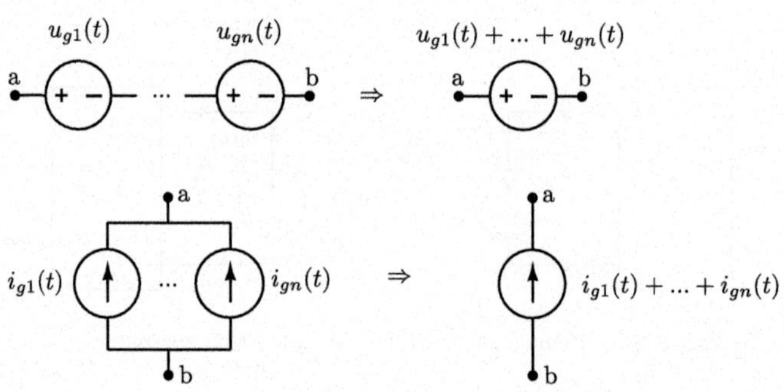

Fig. 1.13 Equivalent circuits of voltage sources connected in series and current sources connected in parallel

1.4 Thévenin's Theorem

Thévenin's theorem states that a linear circuit composed of linear passive elements, independent sources, and dependent sources can be replaced (is equivalent) by a real voltage source.

1.4.1 Direct Current Circuits

Consider a circuit consisting of resistors and DC sources as extensive and complex as desired, denoted as CA. Well, Thévenin's theorem establishes that, from terminals a and b, this circuit is equivalent to a real voltage source as shown in Fig. 1.14. The Thévenin voltage, U_{th}, is the voltage between terminals a and b of the original circuit when open-circuited (Fig. 1.15).

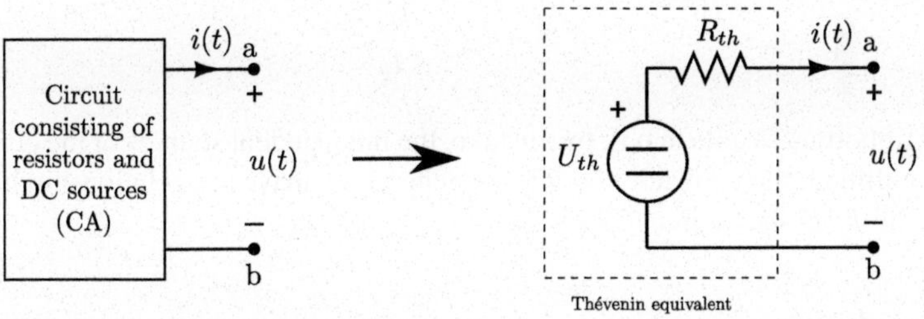

Fig. 1.14 Thévenin equivalent circuit in DC

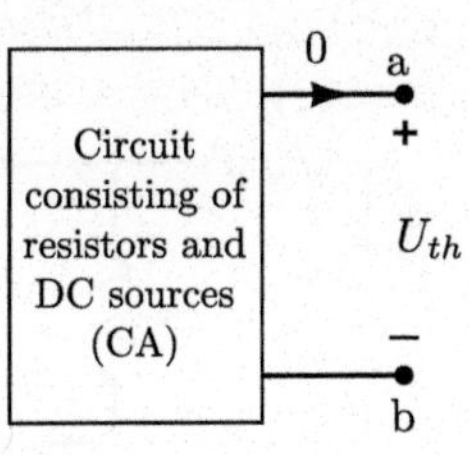

Fig. 1.15 Calculation of Thevenin voltage in DC

The Thévenin resistance, R_{th}, is the resistance obtained between terminals a and b of the original circuit when all independent sources are removed. In this case, there are several ways to calculate the Thévenin resistance:

1. By combining resistances in the passive circuit. This method can only be applied if the circuit does not contain dependent sources.

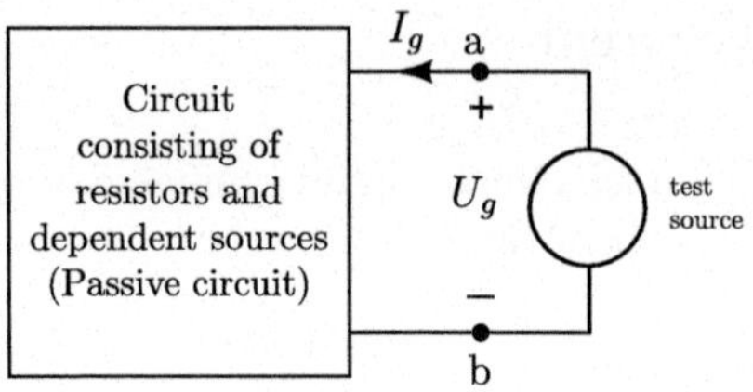

Fig. 1.16 Calculation of Thévenin resistance using a test source

2. By connecting a test source to the passive circuit. This method is used when the circuit has dependent sources, although its use is not limited to this case and can be used in any situation. Figure 1.16 shows the passive circuit with a test source connected between terminals a and b. The test source can be either voltage or current; there are no restrictions on this. The goal is to solve the circuit and obtain the relationship between U_g and I_g. This relationship is precisely the Thévenin resistance of the circuit:

$$R_{th} = \frac{U_g}{I_g} \tag{1.13}$$

It is important to note that if, by mistake, the independent sources of the circuit are not nullified, then it is not possible, in general, to arrive at a relationship between U_g and I_g as shown in (1.13).

Example
Obtain the Thévenin equivalent between terminals a and b of the circuit in Fig. 1.17.

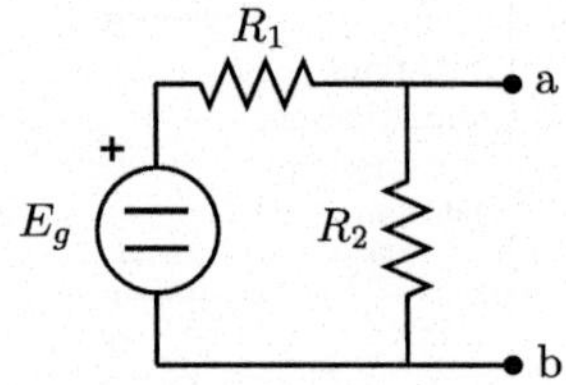

Fig. 1.17 Circuit used in the example under consideration

Solution First, we will find the Thévenin voltage by leaving terminals a and b open as shown in Fig. 1.18.

According to Fig. 1.18, applying the voltage divider concept:

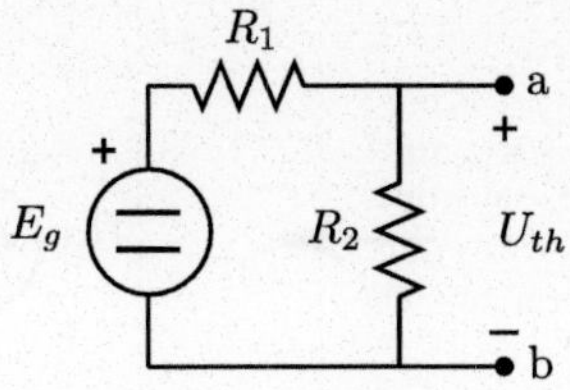

Fig. 1.18 Circuit for the calculation of Thévenin voltage

$$U_{th} = E_g \cdot \frac{R_2}{R_1 + R_2}$$

Next, we will calculate the Thévenin resistance. For this, it is necessary to obtain the passive circuit by nullifying the voltage source as shown in Fig. 1.19.

Referring to Fig. 1.19, it is easy to verify that the two resistors are connected in parallel. Therefore:

$$R_{th} = \frac{R_1 \cdot R_2}{R_1 + R_2}$$

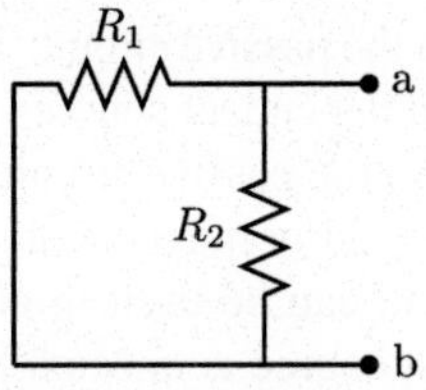

Fig. 1.19 Passive circuit

1.4.2 Alternating Current Circuits

Consider a circuit composed of impedances and AC sources as extensive and complex as desired. Thévenin's theorem states that, from terminals a and b, this circuit is equivalent to a real voltage source as shown in Fig. 1.20.

The Thévenin voltage, $\overline{U}_{th}$, is the voltage between terminals a and b of the original circuit when open-circuited (Fig. 1.21).

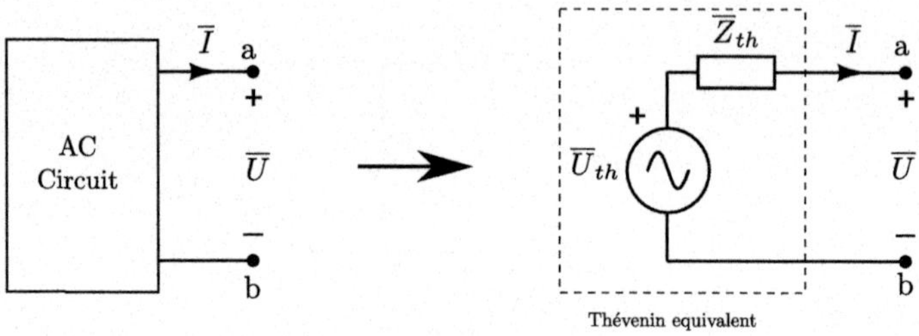

Fig. 1.20 Thévenin equivalent circuit in alternating current

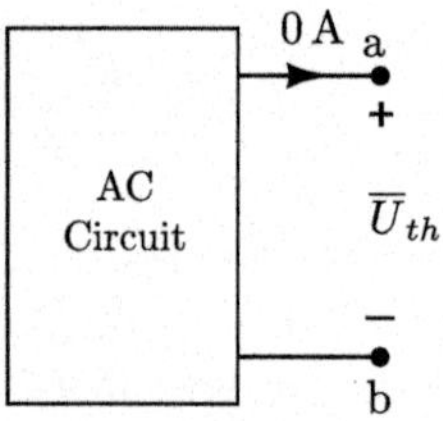

Fig. 1.21 Calculation of Thévenin voltage in alternating current

The Thévenin impedance, $\overline{Z}_{th}$, is the impedance obtained between terminals a and b of the original circuit when all independent sources are removed. In this case, there are several ways to calculate the Thévenin impedance:

1. By combining impedances in the passive circuit. This method can only be applied if the circuit does not contain dependent sources or magnetically coupled coils.
2. By connecting a test source to the passive circuit. This method is used when the circuit has dependent sources and/or magnetically coupled coils, although its use is not limited to this case and can be used in any situation. Figure 1.22 shows the passive circuit with a test source connected between terminals a and b. The test source can be either voltage or current; there are no restrictions on this. The goal is to solve the circuit and obtain the relationship between $\overline{U}_g$ and $\overline{I}_g$. This relationship is precisely the Thévenin impedance of the circuit:

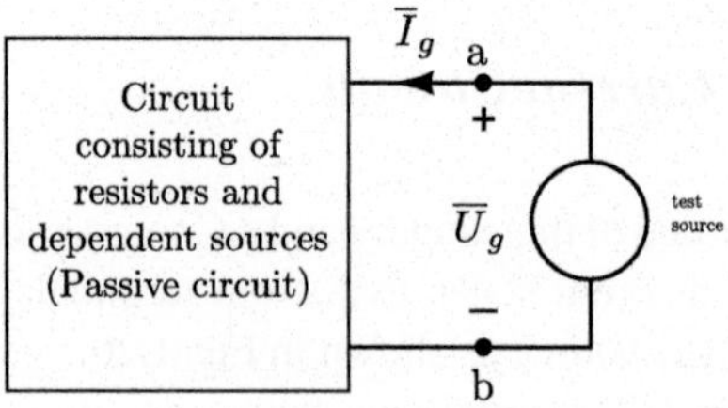

Fig. 1.22 Calculation of Thévenin impedance using a test source

$$\overline{Z}_{th} = \frac{\overline{U}_g}{\overline{I}_g} \tag{1.14}$$

Again, it is important to note that if, by mistake, not all independent sources of the circuit are nullified, then it is not possible, in general, to arrive at a relationship between $\overline{U}_g$ and $\overline{I}_g$ as shown in (1.14).

Example
Obtain the Thévenin equivalent between terminals a and b of the circuit in Fig. 1.23.

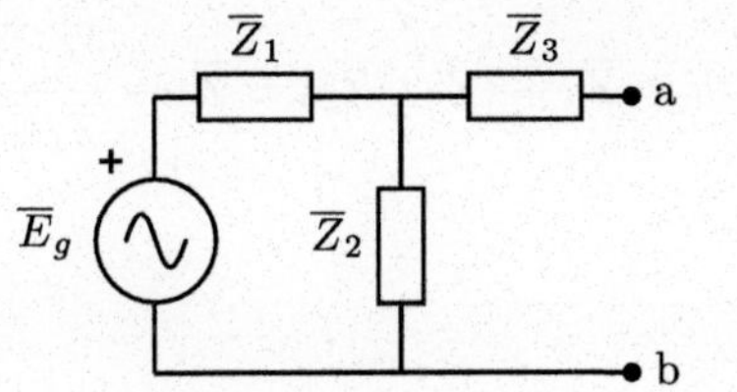

Fig. 1.23 Circuit used in the example under consideration

Solution First, we will find the Thévenin voltage by leaving terminals a and b open as shown in Fig. 1.24.

According to Fig. 1.24, applying the voltage divider concept:

$$\overline{U}_{th} = \overline{E}_g \cdot \frac{\overline{Z}_2}{\overline{Z}_1 + \overline{Z}_2}$$

Next, we will calculate the Thévenin impedance. For this, it is necessary to obtain the passive circuit by nullifying the voltage source as shown in Fig. 1.25.

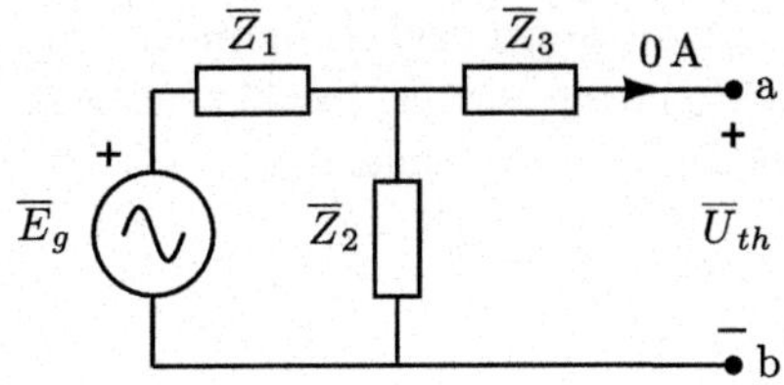

Fig. 1.24 Circuit for the calculation of Thévenin voltage

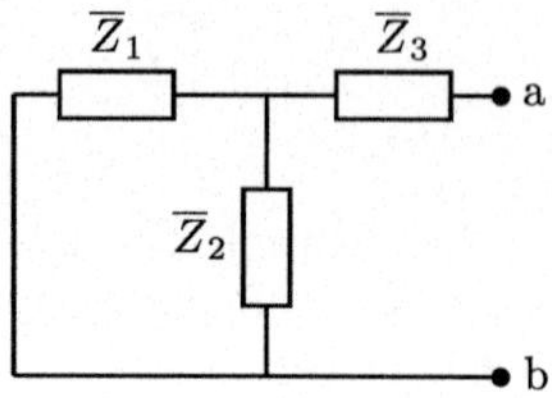

Fig. 1.25 Passive circuit

Referring to Fig. 1.25, it can be observed that impedance $\overline{Z}_3$ is connected in series with the resulting impedance from the parallel combination of $\overline{Z}_1$ and $\overline{Z}_2$:

$$\overline{Z}th = \overline{Z}_3 + (\overline{Z}_1 // \overline{Z}_2) = \overline{Z}_3 + \frac{\overline{Z}_1 \cdot \overline{Z}_2}{\overline{Z}_1 + \overline{Z}_2}$$

1.5 Norton's Theorem

Norton's theorem states that a linear circuit composed of linear passive elements, independent sources, and dependent sources can be replaced (is equivalent) by a real current source.

1.5.1 Direct Current Circuits

Consider a circuit composed of resistors and DC sources as extensive and complex as desired, denoted as CA. Norton's theorem states that, from terminals a and b, this circuit is equivalent to a real current source as shown in Fig. 1.26.

The Norton current, I_{nor}, is the current flowing between terminals a and b when these terminals are short-circuited (Fig. 1.27).

The Norton resistance is obtained between terminals a and b of the original circuit by removing all independent sources. Its calculation procedure is the same as described for the case of Thévenin equivalent.

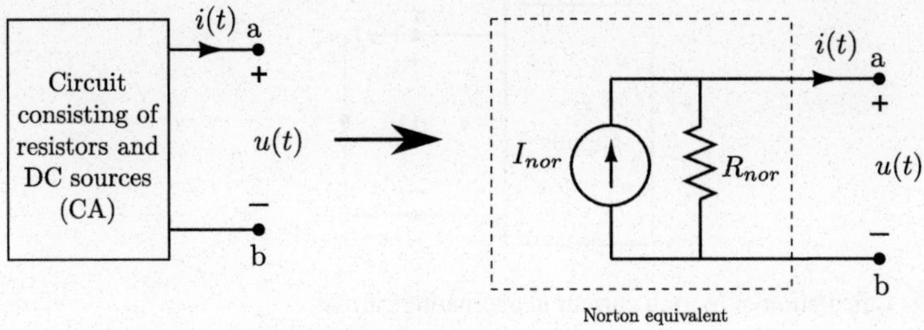

Fig. 1.26 Norton equivalent circuit in direct current

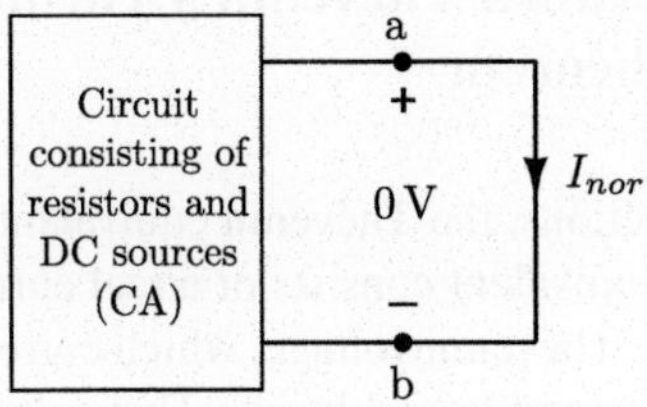

Fig. 1.27 Calculation of Norton current in direct current

1.5.2 Alternating Current Circuits

Consider a circuit composed of impedances and AC sources as extensive and complex as desired. Norton's theorem states that, from terminals a and b, this circuit is equivalent to a real current source, as shown in Fig. 1.28.

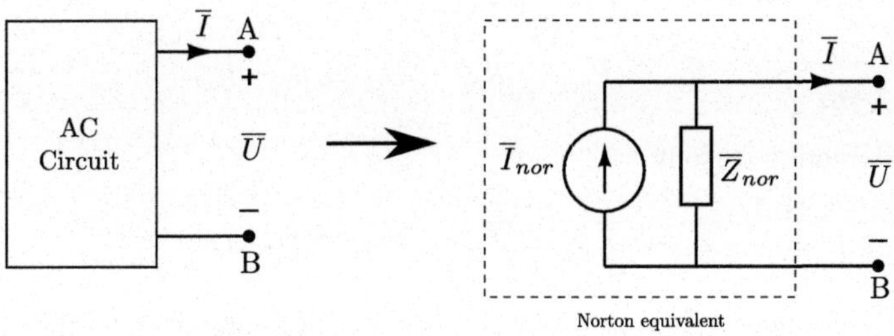

Fig. 1.28 Norton equivalent circuit in alternating current

The Norton current, $\overline{I}_{nor}$, is the current flowing between terminals a and b when these terminals are short-circuited (Fig. 1.29).

The Norton impedance is obtained between terminals a and b of the original circuit by removing all independent sources. Its calculation procedure is the same as described for the case of Thévenin equivalent.

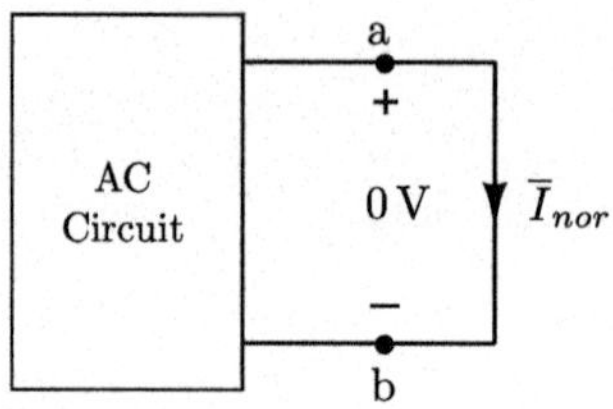

Fig. 1.29 Calculation of Norton current in alternating current

1.6 Relationship Between Thévenin's Theorem and Norton's Theorem

As seen in the previous sections, the Thévenin equivalent consists of a real voltage source, while the Norton equivalent consists of a real current source. The Thévenin equivalent is defined by the Thévenin voltage, which corresponds to the open-circuit voltage between terminals a and b, and by the Thévenin resistance, the equivalent resistance between terminals a and b. On the other hand, the Norton equivalent is defined by the Norton current, which corresponds to the current flowing between terminals a and b once short-circuited, and by the Norton resistance, the equivalent resistance between terminals a and b.

Obtaining both equivalents from the same original circuit implies that they are equivalent to each other. That is, for the same circuit, the Thévenin and Norton equivalents are, in turn, equivalent to each other. This means that the parameters defining them are related to each other. Since both equivalents are real sources, these equivalence relationships will be the same as for real sources, and therefore for the case of direct current they are

$$R_{th} = R_{nor} = R_{eq} \;\; ; \;\; U_{th} = R_{eq} \cdot I_{nor} \tag{1.15}$$

and for alternating current:

$$\overline{Z}_{th} = \overline{Z}_{nor} = \overline{Z}_{eq} \;\; ; \;\; \overline{U}_{th} = \overline{Z}_{eq} \cdot \overline{I}_{nor} \tag{1.16}$$

These equivalence relationships mean that only two parameters of the set formed by Thévenin voltage, Norton current, and equivalent resistance (impedance) are independent. This means that by determining any two of them, the third can be obtained. There are certain problems in which the calculation of Thévenin voltage and Norton current is especially simple, so once both magnitudes are obtained, the resistance (equivalent impedance) can be obtained as the quotient of both:

$$R_{eq} = \frac{U_{th}}{I_{nor}} \;\; ; \;\; \overline{Z}_{eq} = \frac{\overline{U}_{th}}{\overline{I}_{nor}} \tag{1.17}$$

Sometimes, certain circuits allow obtaining the Thévenin or Norton equivalent through real source equivalences.

Example
Obtain the Thévenin and Norton equivalents between terminals a and b of the circuit in Fig. 1.30.

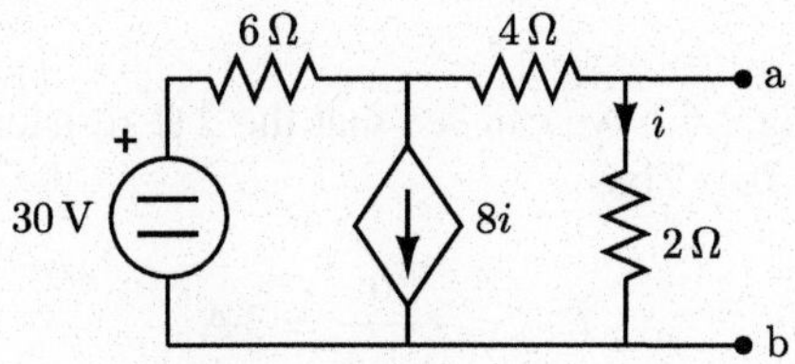

Fig. 1.30 Circuit used in the example under consideration

Solution First, we will find the Thévenin voltage by leaving terminals a and b open as shown in Fig. 1.31.

According to Fig. 1.31:

$$i_1 = 8i + i = 9i$$

Also:

$$30 = 6 \cdot i_1 + (4 + 2) \cdot i = 6 \cdot i_1 + 6 \cdot i = 6 \cdot (9i) + 6 \cdot i = 60i \Rightarrow i = 0{,}5\,\text{A}$$

Consequently:

$$U_{th} = 2 \cdot i = 2 \cdot 0{,}5 = 1\,\text{V}$$

Next, we calculate the Norton current, which is obtained by short-circuiting terminals a and b, as shown in Fig. 1.32.

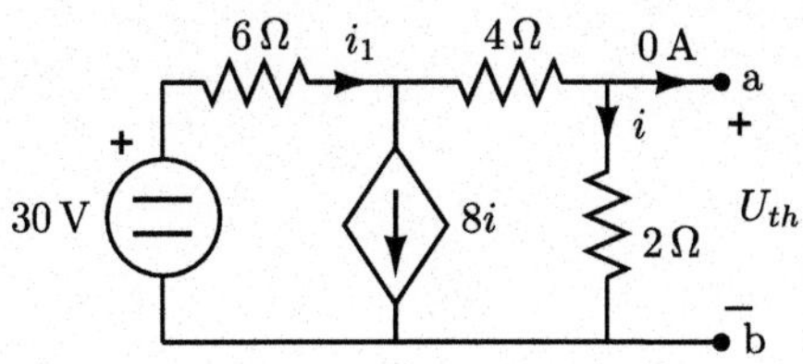

Fig. 1.31 Circuit for the calculation of Thévenin voltage

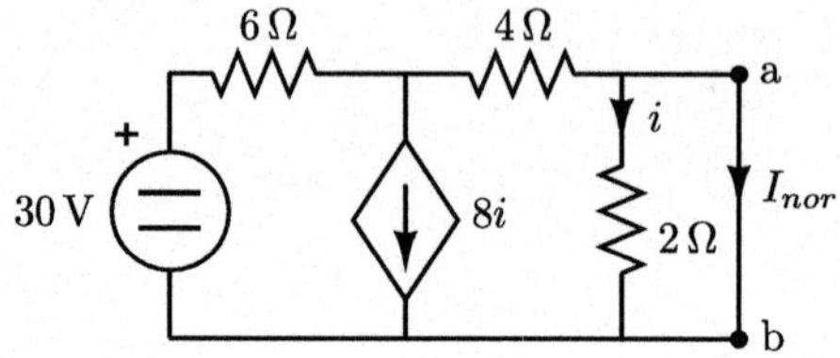

Fig. 1.32 Circuit with terminals a and b short-circuited

According to Fig. 1.32, we can see that the 2 Ω resistor is short-circuited and, therefore, $i = 0$ A. Thus:

$$I_{nor} = \frac{30}{6+4} = 3\,\text{A}$$

Once the Thévenin voltage and Norton current are obtained, the equivalent resistance between terminals a and b is calculated as follows:

$$R_{eq} = \frac{Uth}{Inor} = \frac{1}{3}\,\Omega$$

Next, we will calculate the equivalent resistance in another way. In this case, since the circuit has a dependent source, the equivalent resistance cannot be obtained by combining resistors in the passive circuit. Instead, a test source connected between terminals a and b to the passive circuit can be used, as shown in Fig. 1.33.

The circuit in Fig. 1.33 can be simplified by replacing the dependent current source with its dependent voltage source as shown in Fig. 1.34. According to Fig. 1.34:

$$U_x = (6+4)\cdot(I_x - i) - 48i = 10\cdot I_x - 58i$$

On the other hand:

$$i = \frac{U_x}{2}$$

Therefore:

$$U_x = 10\cdot I_x - 58\cdot\frac{U_x}{2} \;\Rightarrow\; R_{eq} = \frac{U_x}{I_x} = \frac{10}{30} = \frac{1}{3}\,\Omega$$

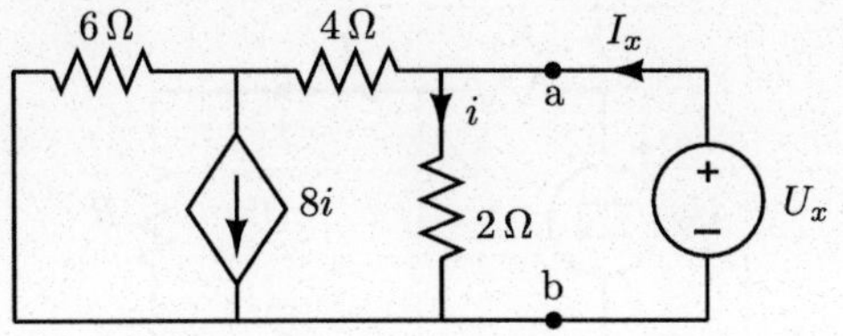

Fig. 1.33 Circuit with a test voltage source

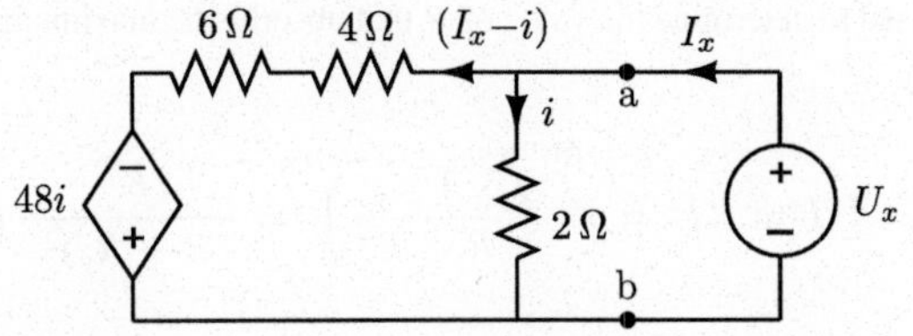

Fig. 1.34 Simplified circuit

1.7 Maximum Power Transfer Theorem

Circuit analysis plays a very important role in the study of systems designed to transfer power between a source and a load. In this regard, there are two main types of systems where emphasis is placed on power transfer. The first type includes power systems, where the emphasis is on the efficiency of power transfer. This is the case for power systems, as their main objective is the efficient generation, transmission, and distribution of electrical energy. In a power system, practically all the power generated at a given moment is absorbed by different loads. The second type of systems focuses on the amount of power transferred. This is the case for communication and instrumentation systems, where the power available at the transmitter or receiver is limited, and therefore, it is desirable for the load to receive the maximum amount of power. In this second type, since the amount of power transmitted is small, the efficiency of this transfer is not a problem.

1.7.1 Maximum Power Transfer from a Real DC Voltage Source

The problem consists of determining the resistance R that absorbs the maximum power from a given linear circuit. In this case, the circuit has been replaced by its Thévenin equivalent, as shown in Fig. 1.35.

According to Fig. 1.35, the power absorbed by the resistance R is as follows:

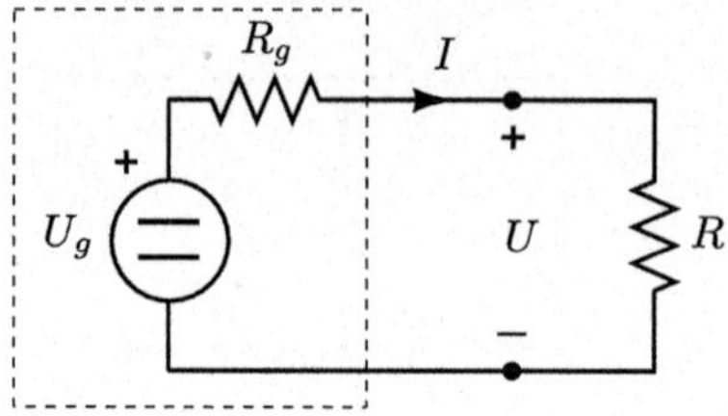

Fig. 1.35 Circuit used to determine the value of R that absorbs the maximum power

$$P = UI = RI^2 = R \cdot \left(\frac{U_g}{R + R_g}\right)^2 = \frac{R}{(R + R_g)^2} \cdot U_g^2 \tag{1.18}$$

It should be noted that U_g and R_g are constants. Therefore, power is only a function of resistance R. To calculate the value of R that maximizes power, we differentiate this function with respect to R:

$$\frac{dP}{dR} = \frac{(R + R_g)^2 - R \cdot 2 \cdot (R + R_g)}{(R + R_g)^4} \cdot U_g^2 \tag{1.19}$$

and setting it to zero gives the value of R:

$$R = R_g \tag{1.20}$$

f Thus, the maximum power transfer occurs when the resistance R is equal to resistance R_g. In this case, the maximum power is as follows:

$$P_{max} = \frac{U_g^2}{4 \cdot R_g} \tag{1.21}$$

On the other hand, the efficiency of the real source is as follows:

$$\eta = \frac{UI}{U_g I} = \frac{R}{R + R_g} \tag{1.22}$$

It can be verified that for the conditions of maximum power transfer, R=R_g, the efficiency is 0,5.

In Fig. 1.36, the variation of the power absorbed by resistance R and the efficiency of the real source as a function of the value of R is represented.

It can be observed that when resistance R is adjusted to receive the maximum power, the efficiency is low.

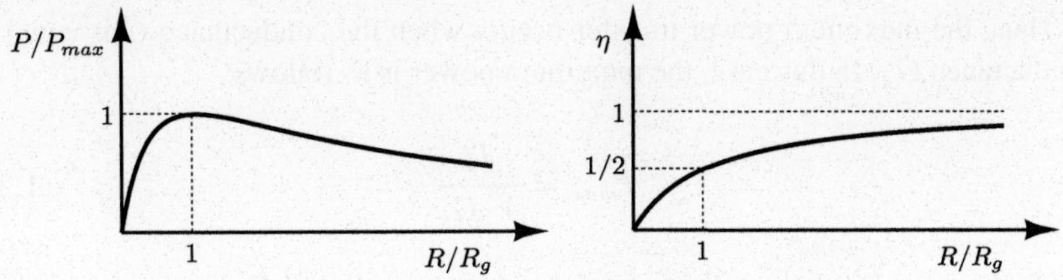

Fig. 1.36 Power and efficiency as a function of the value of R

1.7.2 Maximum Power Transfer from a Real DC Current Source

Now the problem consists of determining the conductance G that absorbs the maximum power from a given linear circuit, modeled by its Norton equivalent as shown in Fig. 1.37.

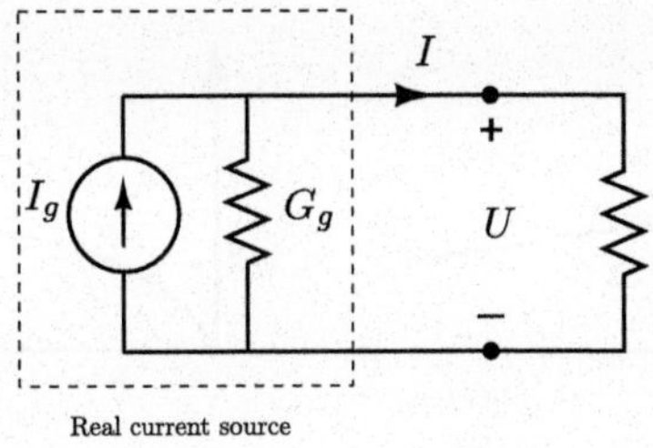

Fig. 1.37 Circuit used to determine the value of G that absorbs the maximum power

According to Fig. 1.37, the power absorbed by conductance G is as follows:

$$P = UI = GU^2 = G \cdot \left(\frac{I_g}{G + G_g} \right)^2 = \frac{G}{(G + G_g)^2} \cdot I_g^2 \tag{1.23}$$

It should be noted that I_g and G_g are constants. Therefore, power is only a function of conductance G. To calculate the value of G that maximizes power, we differentiate this function with respect to G:

$$\frac{dP}{dG} = \frac{(G + G_g)^2 - G \cdot 2 \cdot (G + G_g)}{(G + G_g)^4} \cdot I_g^2 \tag{1.24}$$

and setting it to zero gives the value of G:

$$G = G_g \tag{1.25}$$

Thus, the maximum power transfer occurs when the conductance G is equal to conductance G_g. In this case, the maximum power is as follows:

$$P_{max} = \frac{I_g^2}{4 \cdot G_g} \tag{1.26}$$

On the other hand, the efficiency of the real source is as follows:

$$\eta = \frac{UI}{UI_g} = \frac{G}{G + G_g} \tag{1.27}$$

It can be verified that for the conditions of maximum power transfer, $G{=}G_g$, the efficiency is $0, 5$.

In Fig. 1.38, the variation of the power absorbed by conductance G and the efficiency of the real source as a function of the value of G is represented. It can be observed that when conductance G is adjusted to receive the maximum power, the efficiency is low.

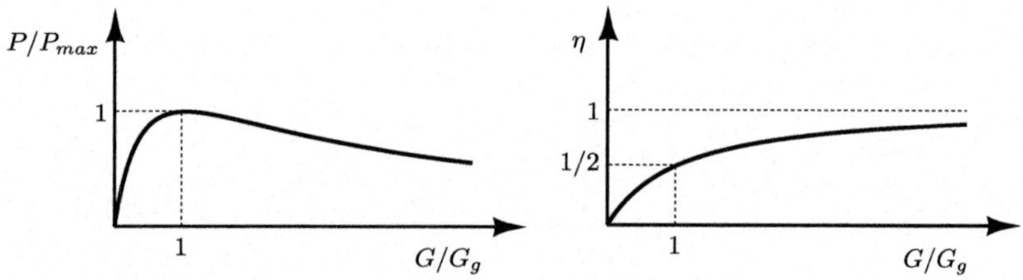

Fig. 1.38 Power and efficiency as a function of the value of G

1.7.3 Maximum Power Transfer from a Real AC Voltage Source

In this case, the goal is to determine the value of the impedance $\overline{Z}$ to absorb the maximum active power from the real voltage source modeling a linear network as shown in Fig. 1.39.

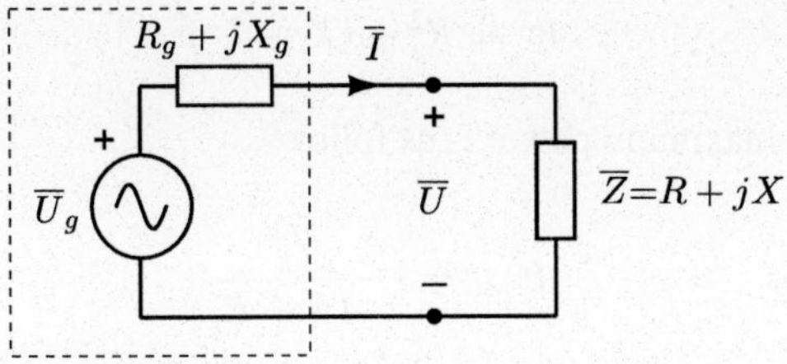

Fig. 1.39 Circuit used to determine the value of $\overline{Z}$ that absorbs the maximum active power

According to Fig. 1.39, the active power absorbed by resistance R is as follows:

$$P = RI^2 = R \cdot \left(\frac{U_g}{\sqrt{(R + R_g)^2 + (X + X_g)^2}} \right)^2 = \frac{R \cdot U_g^2}{(R + R_g)^2 + (X + X_g)^2} \tag{1.28}$$

It should be noted that U_g, R_g, and X_g are constants. In this case, the active power is a function of R and X. To calculate the values of R and X that maximize this power, we need to consider the following three situations:

1. When R is constant and X is variable.
   In this case, the active power depends only on variable X, so to calculate its maximum, we differentiate this function with respect to X:

$$\frac{dP}{dX} = \frac{-2 \cdot R \cdot U_g^2 \cdot (X + X_g)}{\left[(R + R_g)^2 + (X + X_g)^2\right]^2} \tag{1.29}$$

   and setting it to zero gives the value of X:

$$X = -X_g \tag{1.30}$$

   In this case, the maximum power is as follows:

$$P_{max} = \frac{R \cdot U_g^2}{(R + R_g)^2} \tag{1.31}$$

2. When R is variable and X is constant.
   In this case, the active power depends only on variable R, so to calculate its maximum, we differentiate this function with respect to R:

$$\frac{dP}{dR} = \frac{U_g^2 \cdot \left[(R + R_g)^2 + (X + X_g)^2\right] - 2 \cdot R \cdot U_g^2 \cdot (R + R_g)}{\left[(R + R_g)^2 + (X + X_g)^2\right]^2} \tag{1.32}$$

   and setting it to zero gives the value of R:

$$R^2 = R_g^2 + (X + X_g)^2 \tag{1.33}$$

In this case, the maximum power is as follows:

$$P_{max} = \frac{U_g^2}{2 \cdot (R + R_g)} \tag{1.34}$$

3. When both R and X are variable.
   In this case, since both R and X are variable, the active power is a function of both variables. Therefore, to calculate its maximum, we obtain the partial derivatives with respect to each of the variables:

$$\begin{aligned} \frac{\partial P}{\partial X} &= \frac{-2 \cdot R \cdot U_g^2 \cdot (X + X_g)}{\left[(R + R_g)^2 + (X + X_g)^2\right]^2} \\ \frac{\partial P}{\partial R} &= \frac{U_g^2 \cdot \left[(R + R_g)^2 + (X + X_g)^2\right] - 2 \cdot R \cdot U_g^2 \cdot (R + R_g)}{\left[(R + R_g)^2 + (X + X_g)^2\right]^2} \end{aligned} \tag{1.35}$$

   and setting them to zero gives the values of R and X:

$$R = R_g \;\; ; \;\; X = -X_g \tag{1.36}$$

   It can be observed that in the case where both R and X are variables, the maximum power transfer is achieved when the load impedance is equal to the complex conjugate of the source impedance:

$$\overline{Z} = R + jX = \overline{Z}_g^* = R_g - jX_g \tag{1.37}$$

   In this case, the maximum power is as follows:

$$P_{max} = \frac{U_g^2}{4 \cdot R_g} \tag{1.38}$$

   It can be verified that when both R and X are variable, the efficiency is 50%. Among the three cases, the power absorbed by the load in case 3 is the highest of all.

1.7.4 Maximum Power Transfer from a Real AC Current Source

In this case, the goal is to determine the value of the admittance $\overline{Y}$ to absorb the maximum active power from the real current source modeling a linear network as shown in Fig. 1.40.

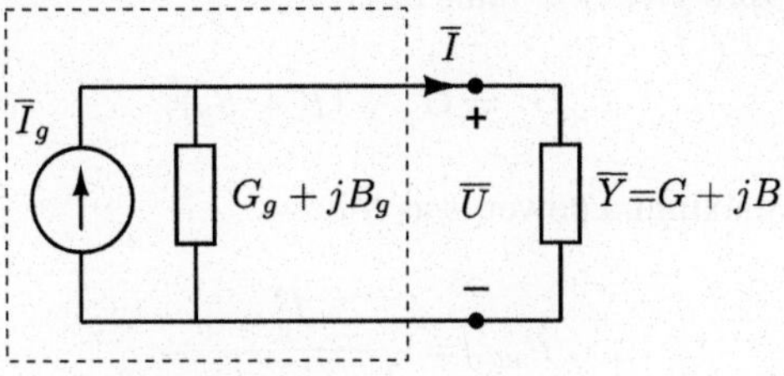

Fig. 1.40 Circuit used to determine the value of $\overline{Y}$ that absorbs the maximum active power

According to Fig. 1.40, the active power absorbed by the conductance G is as follows:

$$P = GU^2 = G \cdot \left(\frac{I_g}{\sqrt{(G+G_g)^2 + (B+B_g)^2}} \right)^2 = \frac{G \cdot I_g^2}{(G+G_g)^2 + (B+B_g)^2} \tag{1.39}$$

It should be noted that I_g, G_g, and B_g are constants. In this case, the active power is a function of G and B. To calculate the values of G and B that maximize this power, we need to consider the following three situations:

1. When G is constant and B is variable.
   In this case, the active power depends only on variable B, so to calculate its maximum, we differentiate this function with respect to B:

$$\frac{dP}{dB} = \frac{-2 \cdot G \cdot I_g^2 \cdot (B+B_g)}{\left[(G+G_g)^2 + (B+B_g)^2\right]^2} \tag{1.40}$$

   and setting it to zero gives the value of B:

$$B = -B_g \tag{1.41}$$

   In this case, the maximum power is as follows:

$$P_{max} = \frac{G \cdot I_g^2}{(G+G_g)^2} \tag{1.42}$$

2. When G is variable and B is constant.
   In this case, the active power depends only on variable G, so to calculate its maximum, we differentiate this function with respect to G:

$$\frac{dP}{dG} = \frac{I_g^2 \cdot \left[(G + G_g)^2 + (B + B_g)^2\right] - 2 \cdot G \cdot I_g^2 \cdot (G + G_g)}{\left[(G + G_g)^2 + (B + B_g)^2\right]^2} \tag{1.43}$$

   and setting it to zero gives the value of G:

$$G^2 = G_g^2 + (B + B_g)^2 \tag{1.44}$$

   In this case, the maximum power is as follows:

$$P_{max} = \frac{I_g^2}{2 \cdot (G + G_g)} \tag{1.45}$$

3. When both G and B are variable.
   In this case, since both G and B are variable, the active power is a function of both variables. Therefore, to calculate its maximum, we obtain the partial derivatives with respect to each of the variables:

$$\begin{aligned} \frac{\partial P}{\partial B} &= \frac{-2 \cdot G \cdot I_g^2 \cdot (B + B_g)}{\left[(G + G_g)^2 + (B + B_g)^2\right]^2} \\ \frac{\partial P}{\partial G} &= \frac{I_g^2 \cdot \left[(G + G_g)^2 + (B + B_g)^2\right] - 2 \cdot G \cdot I_g^2 \cdot (G + G_g)}{\left[(G + G_g)^2 + (B + B_g)^2\right]^2} \end{aligned} \tag{1.46}$$

   and setting them to zero gives the values of G and B:

$$G = G_g \;\; ; \;\; B = -B_g \tag{1.47}$$

   It can be observed that in the case where both G and B are variables, the maximum power transfer is achieved when the load admittance is equal to the complex conjugate of the source admittance:

$$\overline{Y} = G + jB = \overline{Y}_g^* = G_g - jB_g \tag{1.48}$$

   In this case, the maximum power is as follows:

$$P_{max} = \frac{I_g^2}{4 \cdot G_g} \tag{1.49}$$

It can be verified that, of the three cases, the power absorbed by the load in case 3 is the highest of all.

1.8 Coupled Coil Model

Next, we will present the equivalent model of two coupled coils in the phasor domain, which will be used in problem-solving. Although not directly related to the theorems themselves, it has been deemed convenient to introduce this section in order to facilitate the understanding of problems involving coupled coils.

In Fig. 1.41, two coupled coils with the corresponding voltage and current references in the time domain are depicted.

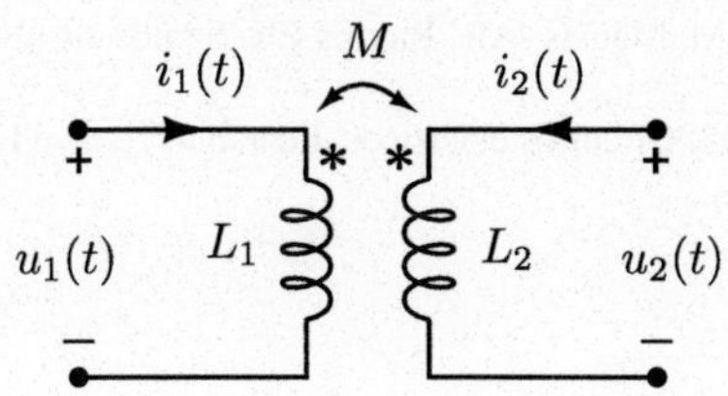

Fig. 1.41 Coupled coils in the time domain

According to Fig. 1.41, the relationship between voltage and current is as follows:

$$
\begin{aligned}
u_1(t) &= L_1 \cdot \frac{di_1(t)}{dt} + M \cdot \frac{di_2(t)}{dt} \\
u_2(t) &= M \cdot \frac{di_1(t)}{dt} + L_2 \cdot \frac{di_2(t)}{dt}
\end{aligned}
\tag{1.50}
$$

In the phasor domain, Eq. (1.50) become:

$$
\begin{aligned}
\overline{U}_1 &= jL_1\omega \cdot \overline{I}_1 + jM\omega \cdot \overline{I}_2 \\
\overline{U}_2 &= jM\omega \cdot \overline{I}_1 + jL_2\omega \cdot \overline{I}_2
\end{aligned}
\tag{1.51}
$$

From Eq. (1.51), the equivalent circuit of two coupled coils in the phasor domain is obtained, as shown in Fig. 1.42.

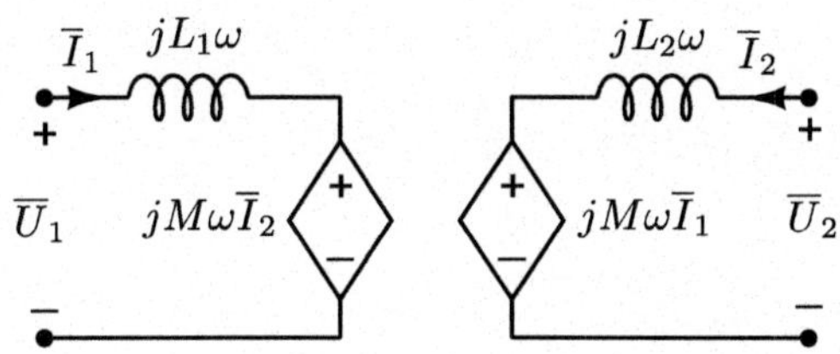

Fig. 1.42 Equivalent model of two coupled coils in the phasor domain

Observing Fig. 1.42, it can be seen that dependent voltage sources have been used to model the coupling between both coils. Similarly, dependent current sources could have been used. However, in this book, only the equivalent model of coupled coils with dependent voltage sources has been considered.

References

1. Alexander CK, Sadiku MNO (2018) Fundamentos de circuitos eléctricos. McGrawHill
2. Expósito AG, Ramos JLM, Macías JAR, Ramos ER, Santos JR (2007) Fundamentos de teoría de circuitos. Thomson
3. Nilson W, Riedel A (2005) Circuitos eléctricos. Pearson-Prentice Hall

Chapter 2
Solved Problems

2.1 Superposition

In the circuit of Fig. 2.1, calculate the current I applying superposition.

Solution Applying superposition, the circuit of Fig. 2.1 can be represented as the sum of the two circuits shown in Fig. 2.2.

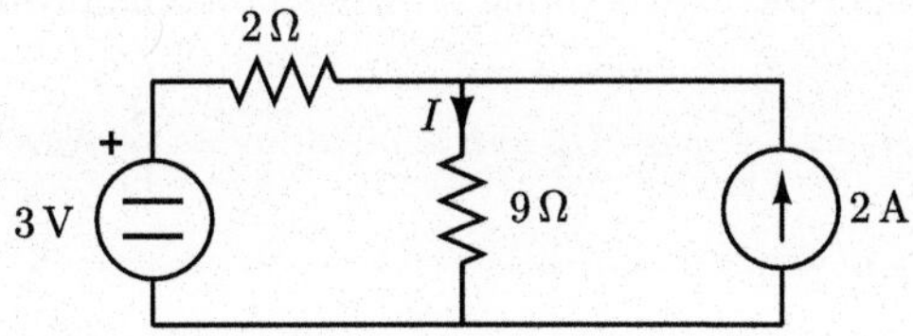

Fig. 2.1 Circuit considered in the problem under study

Considering circuit 1 of Fig. 2.2:

$$I_1 = \frac{3}{2+9} = \frac{3}{11}\,\text{A}$$

On the other hand, in circuit 2 of Fig. 2.2, the concept of current divider can be applied, resulting in:

$$I_2 = 2 \cdot \frac{1/9}{1/2 + 1/9} = \frac{4}{11}\,\text{A}$$

Therefore, the current I is as follows:

$$I = I_1 + I_2 = \frac{3}{11} + \frac{4}{11} = \frac{7}{11}\,\text{A}.$$

A. Bachiller Soler et al., *Theorems for Electrical Circuits*, Lecture Notes in Electrical Engineering 485, https://doi.org/10.1007/978-3-031-74909-4_2

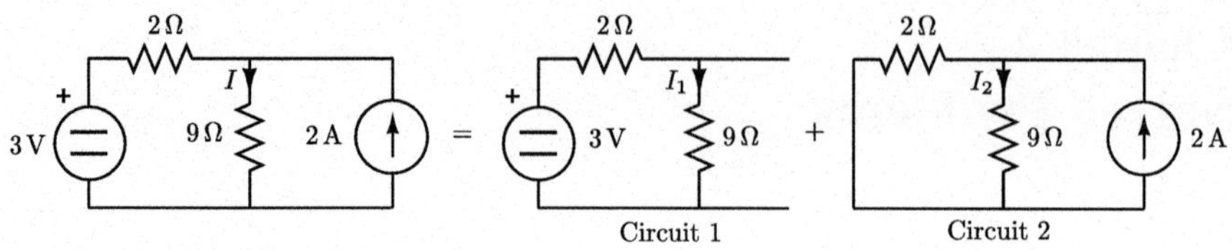

Fig. 2.2 Division of the original circuit

2.2 Proportionality and Superposition

Calculate the new value of the current I of the circuit in Fig. 2.1 if the following changes occur (always referring to the original circuit):

1. The polarity of the voltage source is reversed.
2. The voltage source becomes 6 V and the current source 4 A.
3. The voltage source becomes 6 V and the current source 8 A.

Solution According to the circuit in Fig. 2.1, the current I can be expressed as follows:

$$I = \alpha \cdot U_g + \beta \cdot I_g$$

Taking into account the results of the previous problem, in which superposition was applied:

$$I_1 = \frac{3}{11} = \alpha \cdot U_g = \alpha \cdot 3 \;\Rightarrow\; \alpha = \frac{1}{11}\ \text{A/V}$$

On the other hand:

$$I_2 = \frac{4}{11} = \beta \cdot I_g = \beta \cdot 2 \;\Rightarrow\; \beta = \frac{2}{11}$$

Thus:

$$I = \frac{1}{11} \cdot U_g + \frac{2}{11} \cdot I_g$$

1. If the polarity of the voltage source is reversed, then the current I is as follows:

$$I = \frac{1}{11} \cdot U_g + \frac{2}{11} \cdot I_g = \frac{1}{11} \cdot (-3) + \frac{2}{11} \cdot 2 = \frac{1}{11}\ \text{A}$$

2. If the voltage source becomes 6 V and the current source 4 A, then the current I is as follows:

$$I = \frac{1}{11} \cdot U_g + \frac{2}{11} \cdot I_g = \frac{1}{11} \cdot 6 + \frac{2}{11} \cdot 4 = \frac{14}{11}\ \text{A}$$

3. If the voltage source becomes 6 V and the current source 8 A, then the current I is as follows:

$$I = \frac{1}{11} \cdot U_g + \frac{2}{11} \cdot I_g = \frac{1}{11} \cdot 6 + \frac{2}{11} \cdot 8 = \frac{22}{11} \text{ A}.$$

2.3 Superposition

Apply the superposition theorem to determine the voltage U_x in the circuit of Fig. 2.3.

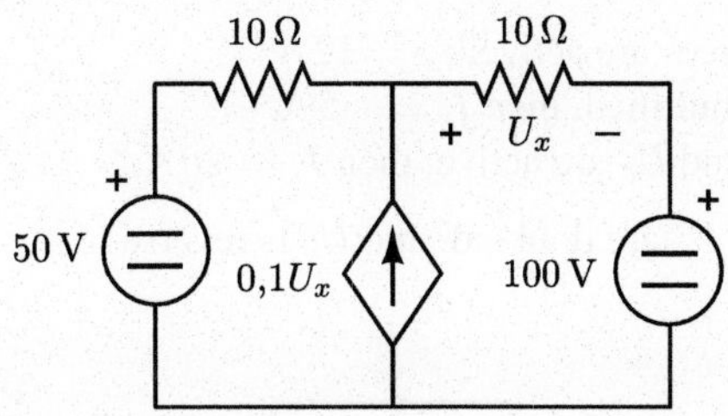

Fig. 2.3 Circuit considered in the problem under study

Solution Applying superposition, the circuit in Fig. 2.3 can be represented as the sum of the two circuits shown in Fig. 2.4. It can be observed that the current-dependent source is present in both circuits, as only the independent sources are superimposed.

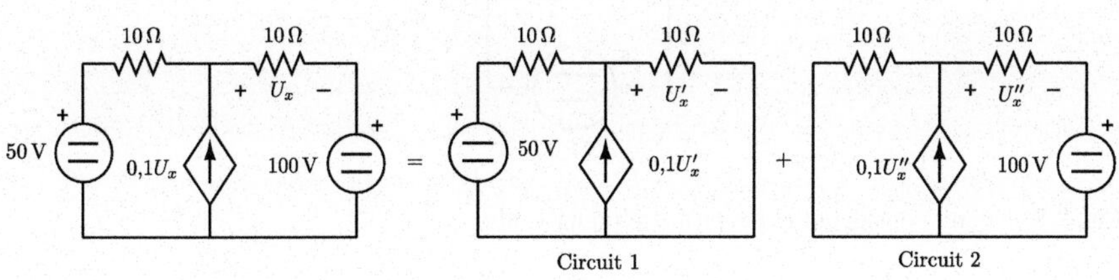

Fig. 2.4 Division of the original circuit

Considering circuit 1 of Fig. 2.4, the nodal analysis method can be applied, after replacing the real voltage source with its equivalent real current source, resulting in:

$$\left(\frac{1}{10} + \frac{1}{10}\right) \cdot U'_x = \frac{50}{10} + 0{,}1U'_x \;\Rightarrow\; U'_x = 50 \text{ V}$$

On the other hand, applying the nodal analysis method again to circuit 2, after replacing the real voltage source with its equivalent real current source, results in:

$$\left(\frac{1}{10}+\frac{1}{10}\right)\cdot(U_x''+100)=\frac{100}{10}+0,1U_x'' \Rightarrow U_x''=-100\,\mathrm{V}$$

Consequently, the voltage U_x is as follows:

$$U_x = U_x' + U_x'' = 50 - 100 = -50\,\mathrm{V}.$$

2.4 Proportionality and Superposition

The circuit C.P. in Fig. 2.5 does not contain independent sources. It is known that:

- When all three sources are active, $I_x = 12\,\mathrm{A}$.
- If the source U_b is nullified, then $I_x = -5\,\mathrm{A}$.
- If only sources I_a and U_b are active, then $I_x = 20\,\mathrm{A}$.

Calculate I_x if I_a and U_c are doubled, and U_b is reversed.

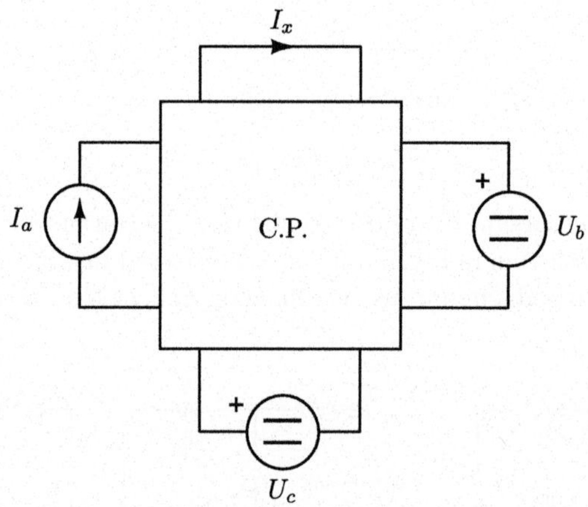

Fig. 2.5 Circuit considered in the problem under study

Solution Before starting the resolution of the problem, it is convenient to make a comment about the nomenclature that will be used:

- I_x^a denotes the current I_x when only the source I_a is active.
- I_x^b denotes the current I_x when only the source U_b is active.
- I_x^c denotes the current I_x when only the source U_c is active.

According to the statement, when all three sources are active, the current I_x is 12 A. Therefore, applying superposition yields:

$$I_x^a + I_x^b + I_x^c = 12 \tag{2.1}$$

If the source U_b is nullified, then the current I_x is -5 A. Therefore, applying superposition yields:

$$I_x^a + I_x^c = -5 \tag{2.2}$$

Lastly, if only sources I_a and U_b are active, then the current I_x is 20 A. Therefore, applying superposition yields:

$$I_x^a + I_x^b = 20 \tag{2.3}$$

Solving the system of three equations with three unknowns (2.1), (2.2), and (2.3) results in:

$$I_x^a = 3\,\text{A} \;;\quad I_x^b = 17\,\text{A} \;;\quad I_x^c = -8\,\text{A}$$

When the values of I_a and U_c are doubled, and U_b is reversed, then I_x is calculated applying proportionality and superposition as follows:

$$I_x' = 2 \cdot I_x^a - I_x^b + 2 \cdot I_x^c = 2 \cdot 3 - 17 - 2 \cdot 8 = -27\,\text{A}.$$

2.5 Superposition

In the circuit of Fig. 2.6, applying superposition, calculate the voltage U_{AB} (voltage between nodes A and B).

Solution Applying superposition, the circuit in Fig. 2.6 can be represented as the sum of the three circuits shown in Fig. 2.7. Considering Circuit 1 in Fig. 2.7, the voltage division concept can be applied resulting in[1]:

$$U'_{AB} = 75 \cdot \frac{30//30}{30//30 + 10} = 75 \cdot \frac{15}{15 + 10} = 45\,\text{V}$$

Next, Circuit 2 will be analyzed, in which the real voltage source will be replaced by its equivalent real current source, as shown in Fig. 2.8.

According to Fig. 2.8, applying the nodal analysis method results in:

$$\left(\frac{1}{10} + \frac{1}{30} + \frac{1}{30}\right) \cdot U''_{AB} = 2 \;\Rightarrow\; U''_{AB} = 12\,\text{V}$$

Lastly, Circuit 3 will be analyzed, in which the real current source will be replaced by its equivalent real voltage source, as shown in Fig. 2.9. Applying the voltage division concept yields the voltage between A and B:

$$U'''_{AB} = 40 \cdot \frac{15/2}{15/2 + 30} = 8\,\text{V}$$

[1] The symbol // means parallel connection.

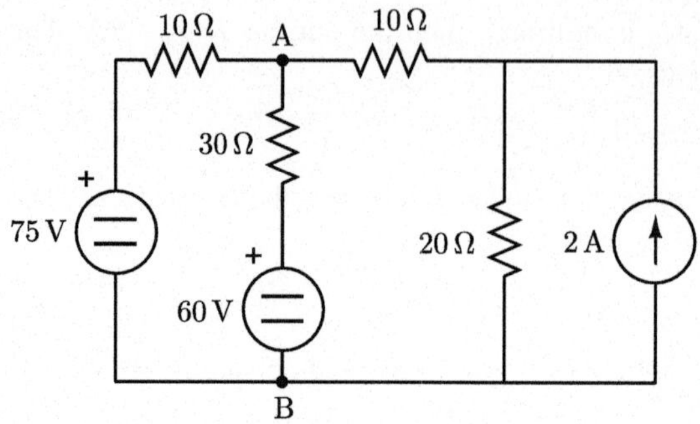

Fig. 2.6 Circuit considered in the problem under study

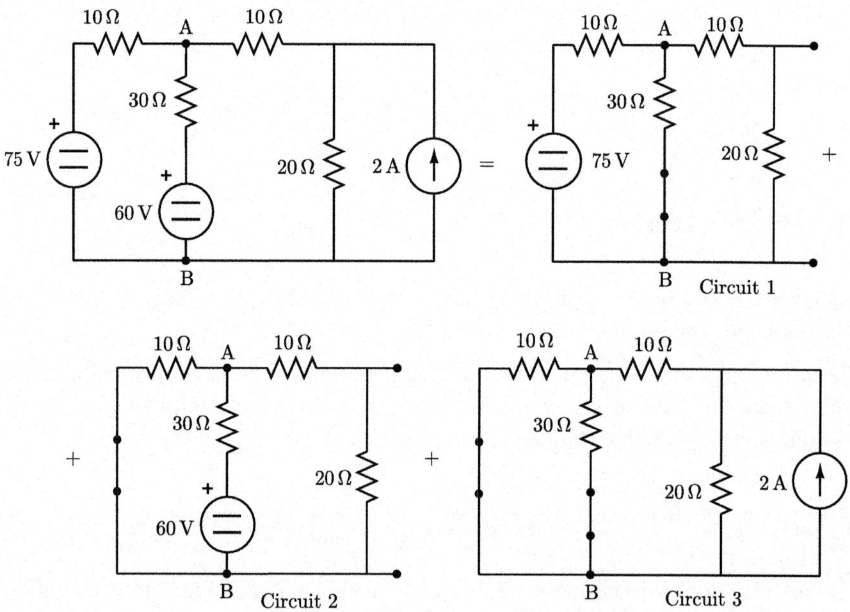

Fig. 2.7 Application of superposition theorem

Therefore, the voltage U_{AB} is as follows:

$$U_{AB} = U'_{AB} + U''_{AB} + U'''_{AB} = 45 + 12 + 8 = 65\,\text{V}.$$

2.6 Superposition

In the circuit of Fig. 2.10, calculate the current $i(t)$ in steady state, knowing that $e_g(t) = 100\sqrt{2}\sin(100t)$ V and $i_g(t) = 6\sqrt{2}\sin(300t)$ A.

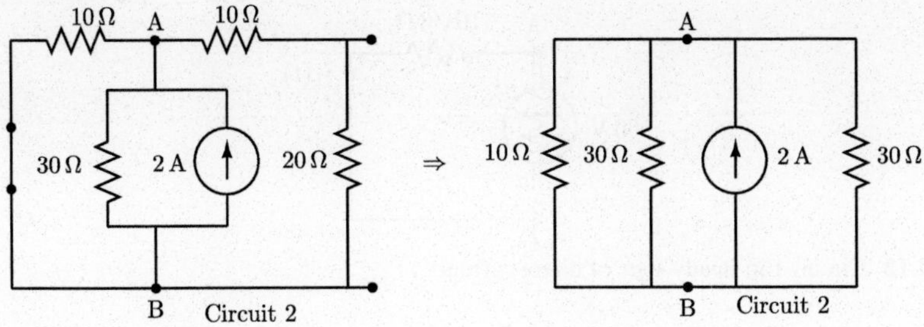

Fig. 2.8 Simplification of circuit 2

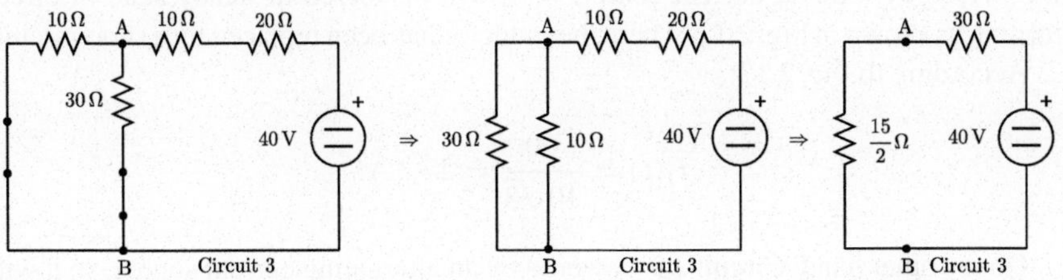

Fig. 2.9 Simplification of circuit 3

Solution Applying superposition, the circuit in Fig. 2.10 can be represented as the sum of the three circuits shown in Fig. 2.11.

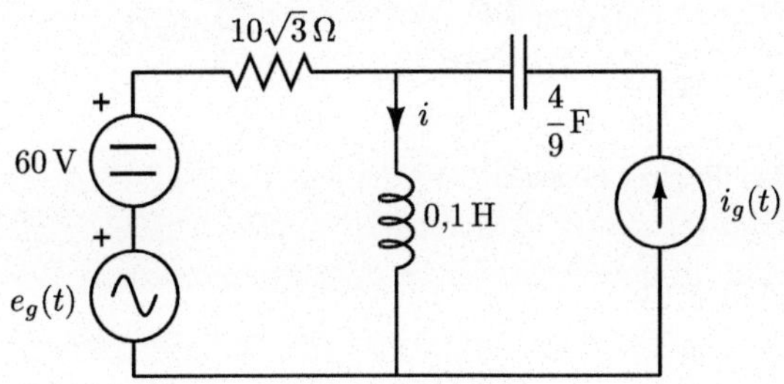

Fig. 2.10 Circuit considered in the problem under study

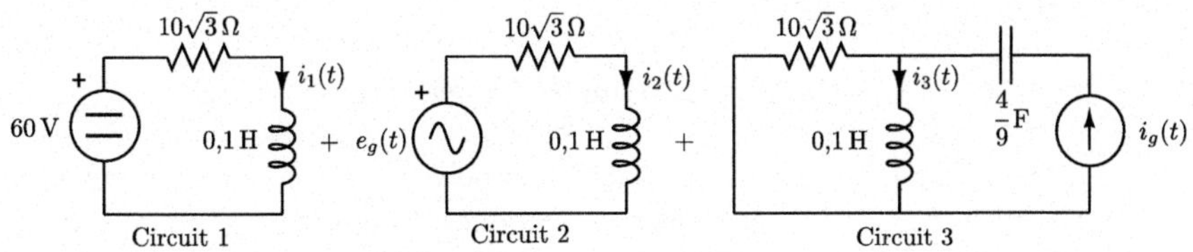

Fig. 2.11 Division of the original circuit

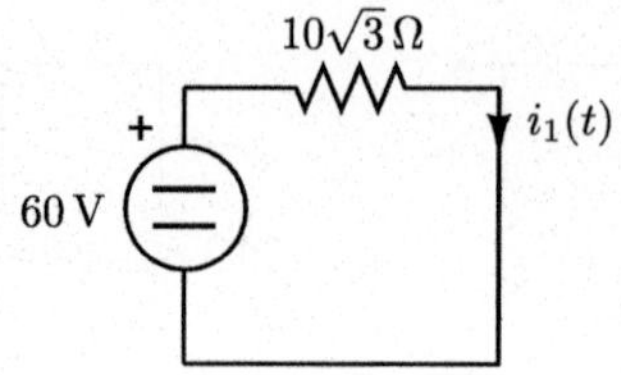

Fig. 2.12 Circuit 1 in steady state of direct current

Next, each of the circuits shown in Fig. 2.11 will be separately solved. Circuit 1 is powered by a direct current source, so it will be solved in steady state of direct current, as shown in Fig. 2.12, where the inductor has been replaced by a short circuit.

According to Fig. 2.12:

$$i_1(t) = \frac{60}{10\sqrt{3}} = 2\sqrt{3}\,\text{A}$$

On the other hand, Circuit 2 is powered by an alternating current source, so it will be solved using phasor techniques. In this sense, Circuit 2 in the phasor domain is as shown in Fig. 2.13.

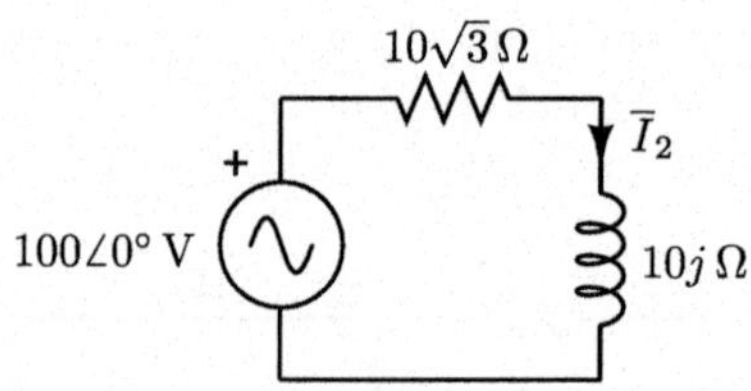

Fig. 2.13 Circuit 2 in the phasor domain

According to Fig. 2.13:

$$\overline{I}_2 = \frac{100\angle 0^\circ}{10\sqrt{3} + 10j} = 5\angle -30^\circ\,\text{A}$$

Once the current $\overline{I}_2$ in the phasor domain is calculated, it is necessary to obtain the associated time function:

$$i_2(t) = 5\sqrt{2}\sin(100t - 30^\circ)\,\text{A}$$

Regarding Circuit 3, it can be seen that it is powered by an alternating current source, so it will be solved using phasor techniques. In this sense, Circuit 3 in the phasor domain is as shown in Fig. 2.14. It can be observed that the value of the impedance of the capacitor has not been represented since it does not intervene in the calculation of $\overline{I}_3$ as it is in series with a current source.

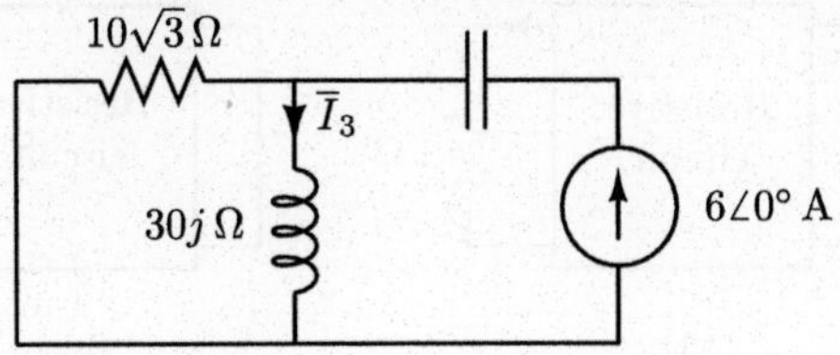

Fig. 2.14 Circuit 3 in the phasor domain

According to Fig. 2.14, applying the concept of current divider results in:

$$\overline{I}_3 = 6\angle 0^\circ \frac{\frac{1}{30j}}{\frac{1}{10\sqrt{3}} + \frac{1}{30j}} = 3\angle -60^\circ \text{ A}$$

Once the current $\overline{I}_3$ in the phasor domain is calculated, it is necessary to obtain the associated time function:

$$i_3(t) = 3\sqrt{2}\sin(300t - 60^\circ) \text{ A}$$

Having obtained the currents $i_1(t)$, $i_2(t)$, and $i_3(t)$, then the current $i(t)$ is as follows:

$$i(t) = i_1(t) + i_2(t) + i_3(t) = 2\sqrt{3} + 5\sqrt{2}\sin(100t - 30^\circ) + 3\sqrt{2}\sin(300t - 60^\circ) \text{ A.}$$

2.7 Proportionality and Superposition

Given the data shown in the circuits of Fig. 2.15, determine the power consumed by the 1 Ω resistor in circuit (c).

Solution Applying superposition, circuit (c) in Fig. 2.15 can be represented as the sum of the two circuits shown in Fig. 2.16. According to this, the current I_R can be obtained as follows:

$$I_R = I_{R1} + I_{R2}$$

From circuit 1 in Fig. 2.16, applying proportionality:

$$I_{R1} = \alpha \cdot 5$$

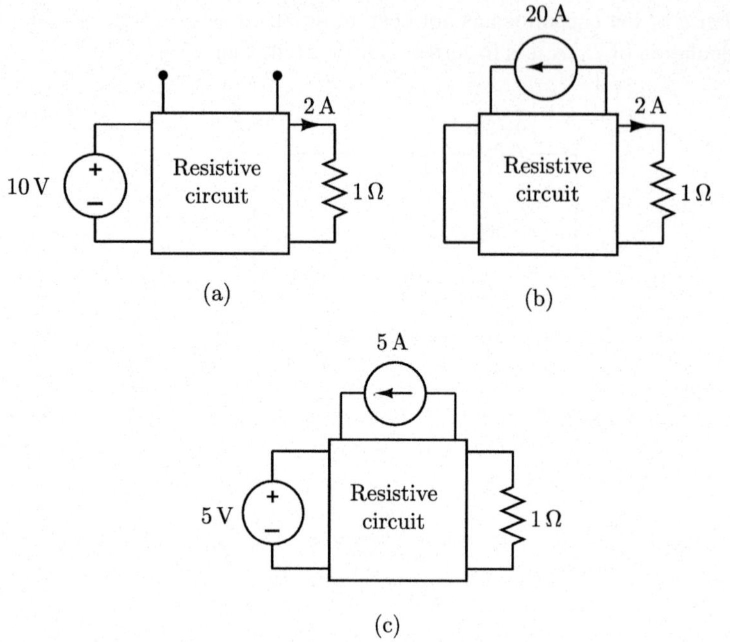

Fig. 2.15 Circuit considered in the problem under study

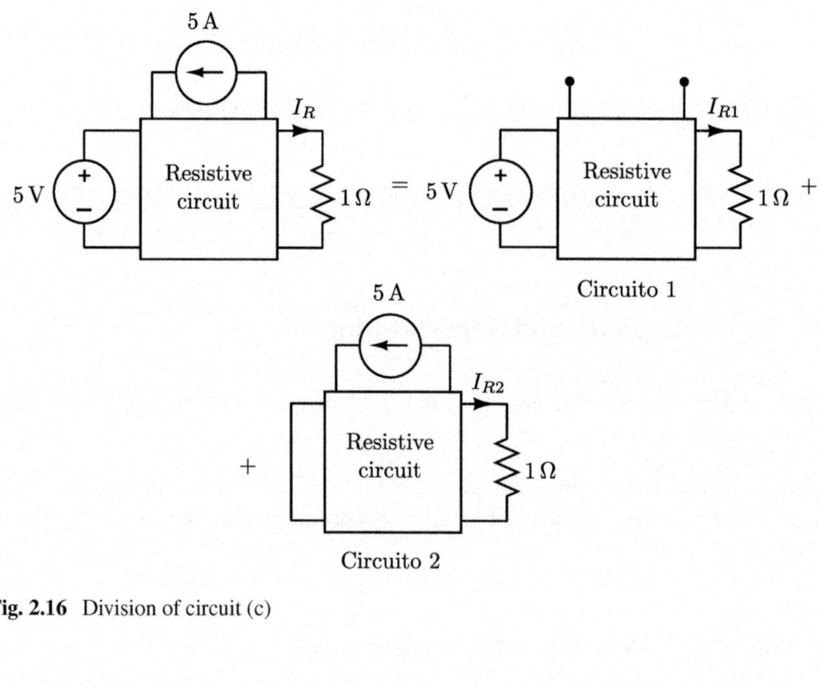

Fig. 2.16 Division of circuit (c)

The coefficient α is obtained from circuit (a) in Fig. 2.15:

$$2 = \alpha \cdot 10 \Rightarrow \alpha = \frac{1}{5} \frac{\text{A}}{\text{V}}$$

Therefore:

$$I_{R1} = \alpha \cdot 5 = \frac{1}{5} \cdot 5 = 1\ \text{A}$$

From circuit 2 in Fig. 2.16, applying proportionality:

$$I_{R2} = \beta \cdot 5$$

The coefficient β is obtained from circuit (b) in Fig. 2.15:

$$2 = \beta \cdot 20 \Rightarrow \beta = \frac{1}{10}$$

Therefore:

$$I_{R2} = \beta \cdot 5 = \frac{1}{10} \cdot 5 = \frac{1}{2}\ \text{A}$$

Thus, the current I_R is as follows:

$$I_R = I_{R1} + I_{R2} = 1 + \frac{1}{2} = \frac{3}{2}\ \text{A}$$

Consequently, the power absorbed by the 1 Ω resistor is:

$$P = R \cdot I_R^2 = 1 \cdot \left(\frac{3}{2}\right)^2 = \frac{9}{4}\ \text{W}.$$

2.8 Proportionality and Superposition

Given the data shown in the circuits of Fig. 2.17, determine the current $i(t)$ flowing through the resistor in circuit (c).

Solution Applying superposition, circuit (c) in Fig. 2.17 can be represented as the sum of the two circuits shown in Fig. 2.18.

According to Fig. 2.18, the current $i(t)$ can be obtained as follows:

$$i(t) = i_1(t) + i_2(t)$$

From circuit 1 in Fig. 2.18, applying proportionality:

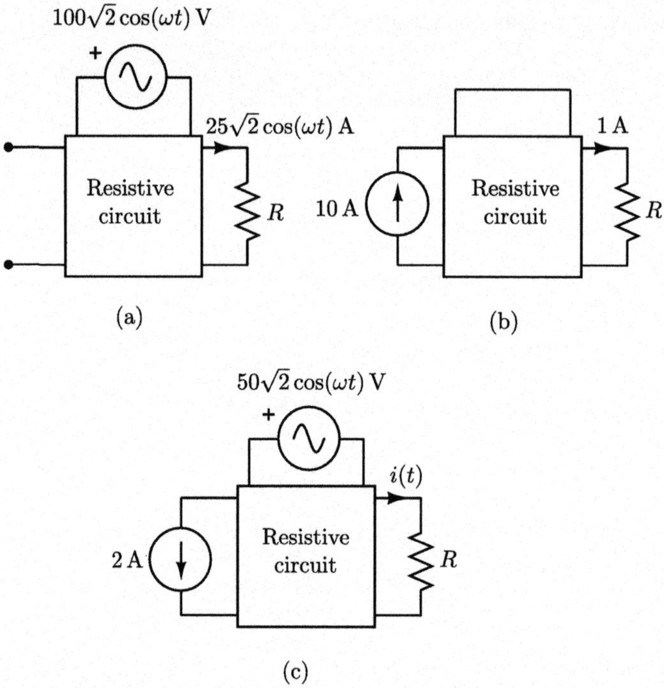

Fig. 2.17 Circuits considered in the problem under study

$$i_1(t) = \alpha \cdot 50\sqrt{2}\cos(\omega t)$$

The coefficient α is obtained from circuit (a) in Fig. 2.17:

$$25\sqrt{2}\cos(\omega t) = \alpha \cdot 100\sqrt{2}\cos(\omega t) \Rightarrow \alpha = \frac{25\sqrt{2}\cos(\omega t)}{100\sqrt{2}\cos(\omega t)} = \frac{1}{4}\frac{\text{A}}{\text{V}}$$

Therefore:

$$i_1(t) = \alpha \cdot 50\sqrt{2}\cos(\omega t) = \frac{1}{4} \cdot 50\sqrt{2}\cos(\omega t)\,\text{A}$$

From circuit 2 in Fig. 2.18, applying proportionality:

$$i_2(t) = \beta \cdot 2$$

The coefficient β is obtained from circuit (b) in Fig. 2.17:

$$1 = \beta \cdot (-10) \Rightarrow \beta = \frac{-1}{10}$$

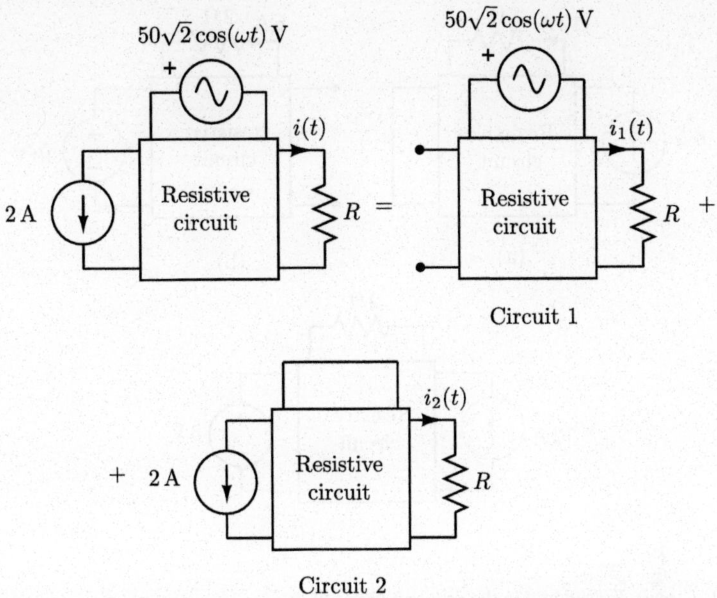

Fig. 2.18 Division of circuit (c)

The negative sign of the current is due to the difference in direction between the current source in circuit (b) of Fig. 2.17 and that in circuit 2 of Fig. 2.18.

Therefore:

$$i_2(t) = \beta \cdot 2 = \frac{-1}{10} \cdot 2 = \frac{-1}{5}\ \text{A}$$

Thus, the current $i(t)$ is as follows:

$$i(t) = i_1(t) + i_2(t) = \frac{1}{4} \cdot 50\sqrt{2}\cos(\omega t) - \frac{1}{5}\ \text{A}.$$

2.9 Proportionality and Superposition

Given the data shown in the following circuits of Fig. 2.19, determine the power consumed by the $2\,\Omega$ resistor in circuit (c).

Solution Applying superposition, circuit (c) in Fig. 2.19 can be represented as the sum of the two circuits shown in Fig. 2.20, according to which, the current I_R can be obtained as follows:

$$I_R = I_{R1} + I_{R2}$$

From circuit 1 in Fig. 2.20, applying proportionality:

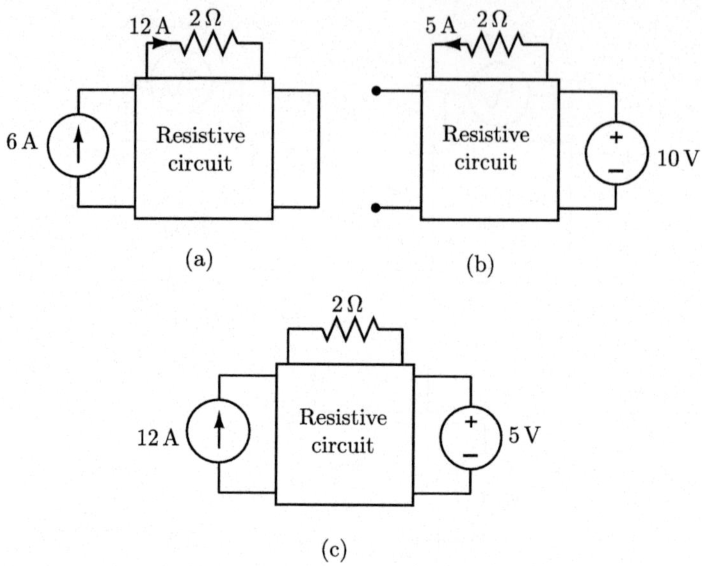

Fig. 2.19 Circuits considered in the problem under study

$$I_{R1} = \alpha \cdot 5$$

The coefficient α is obtained from circuit (b) in Fig. 2.19:

$$-5 = \alpha \cdot 10 \;\Rightarrow\; \alpha = \frac{-1}{2}\frac{\mathrm{A}}{\mathrm{V}}$$

Note that the negative sign is because the direction of the 5 A current source in circuit (b) of Fig. 2.19 is opposite to the direction of the current I_{R1} in circuit 1 of Fig. 2.20.

Therefore:

$$I_{R1} = \alpha \cdot 5 = \frac{-1}{2} \cdot 5 = \frac{-5}{2}\,\mathrm{A}$$

From circuit 2 in Fig. 2.20, applying proportionality:

$$I_{R2} = \beta \cdot 12$$

The coefficient β is obtained from circuit (a) in Fig. 2.19:

$$12 = \beta \cdot 6 \;\Rightarrow\; \beta = 2$$

Therefore:

$$I_{R2} = \beta \cdot 12 = 2 \cdot 12 = 24\,\mathrm{A}$$

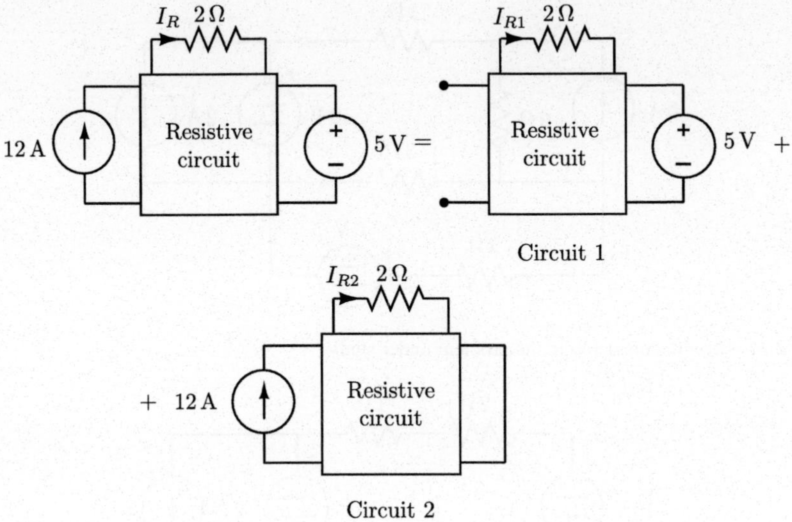

Fig. 2.20 Division of circuit (c)

Thus, the current I_R is as follows:

$$I_R = I_{R1} + I_{R2} = \frac{-5}{2} + 24 = \frac{43}{2}\ \text{A}$$

Consequently, the power absorbed by the 2 Ω resistor is:

$$P = R \cdot I_R^2 = 2 \cdot \left(\frac{43}{2}\right)^2 \approx 924{,}5\ \text{W}.$$

2.10 Equivalents

In the circuit of Fig. 2.21, determine the power of the 5 A source and the 10 V source, indicating whether they are absorbed or supplied. Note: It is recommended to make simplifications and transformations in order to solve the circuit with a single loop.

Solution First, the real current source of 5 A and 2 Ω will be replaced by its equivalent real voltage source, resulting in the circuit of Fig. 2.22.

From the circuit in Fig. 2.22, the real voltage source of 6 V and 2 Ω will be replaced by its equivalent real current source. Likewise, the parallel association of the 10 V voltage source and the 2 A current source is equivalent to a 10 V voltage source, resulting in the circuit in Fig. 2.23.

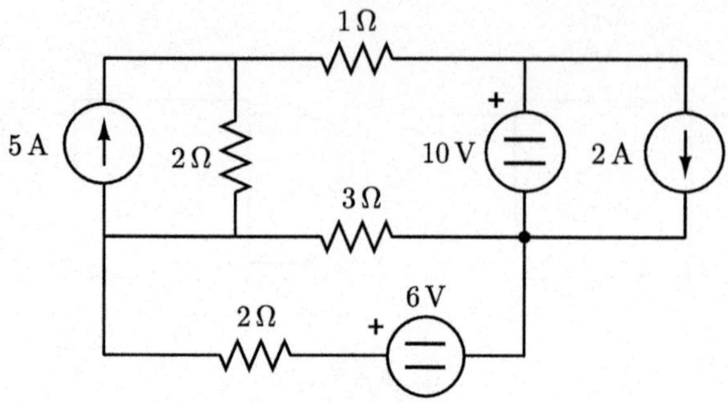

Fig. 2.21 Circuit considered in the problem under study

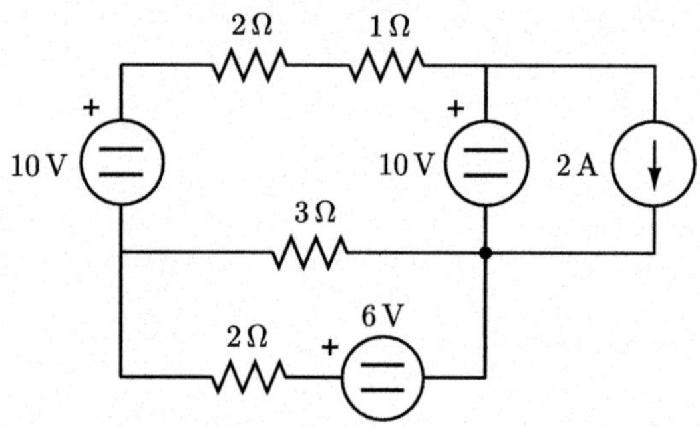

Fig. 2.22 Transformation of the original circuit

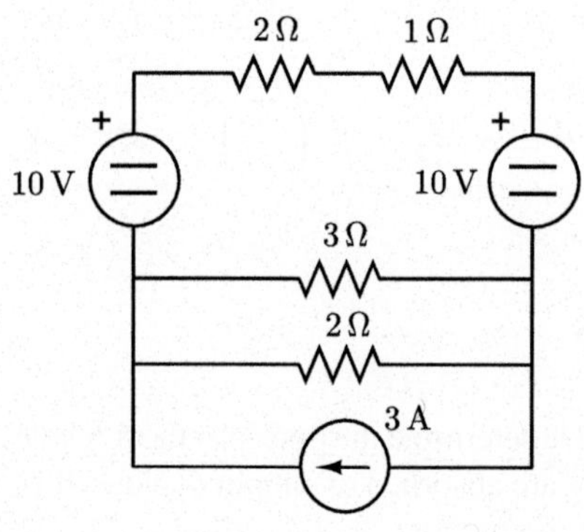

Fig. 2.23 Transformation of the original circuit

Finally, the equivalent resistance of the parallel connection of the 3 Ω and 2 Ω resistors is obtained, and the real current source obtained is replaced by its equivalent real voltage source, as shown in Fig. 2.24.

According to Fig. 2.24:

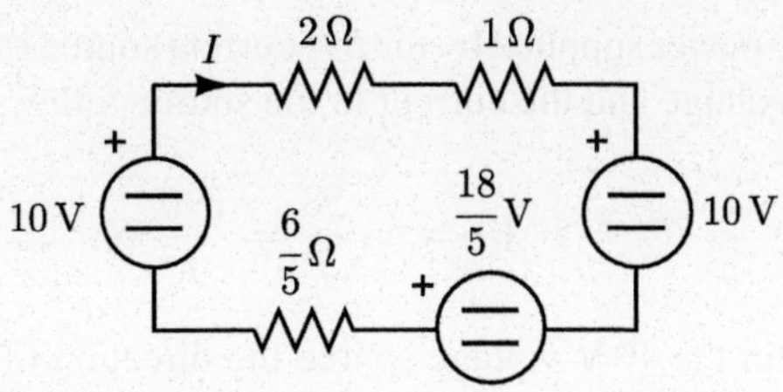

Fig. 2.24 Transformation of the original circuit

$$I = \frac{10 - 10 + \frac{18}{5}}{2 + 1 + \frac{6}{5}} = \frac{6}{7}\,\text{A}$$

Once the current I is obtained, it is necessary to return to the original circuit and calculate the quantities of interest. In particular, the voltage of the 5 A current source, U_5, and the current of the 10 V voltage source, I_{10}, need to be calculated (Fig. 2.25). Note that the 10 V source in Fig. 2.24 is not the same as the 10 V source in the original circuit.

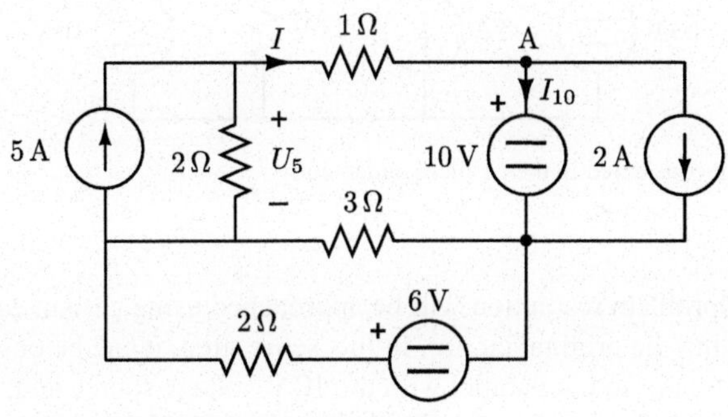

Fig. 2.25 Original circuit in the problem under study

According to Fig. 2.25, applying Kirchhoff's current law at node A yields:

$$I = 2 + I_{10} \;\Rightarrow\; I_{10} = I - 2 = \frac{6}{7} - 2 = \frac{-8}{7}\,\text{A}$$

On the other hand, the voltage U_5 is calculated as follows:

$$U_5 = (5 - I) \cdot 2 = \left(5 - \frac{6}{7}\right) \cdot 2 = \frac{58}{7}\,\text{V}$$

Consequently, the power supplied by the 5 A current source is as follows (note that the direction of the voltage and the current in the source follows an active reference criterion):

$$P_5 = 5 \cdot U_5 = 5 \cdot \frac{58}{7} = \frac{290}{7}\ \mathrm{W}$$

Considering that in the 10 V voltage source the direction of the voltage and the current follows a passive reference criterion, the power absorbed by it is as follows:

$$P_{10} = 10 \cdot I_{10} = 10 \cdot \frac{-8}{7} = \frac{-80}{7}\ \mathrm{W}.$$

2.11 Equivalents

In the circuit of Fig. 2.26, calculate the current I_R.

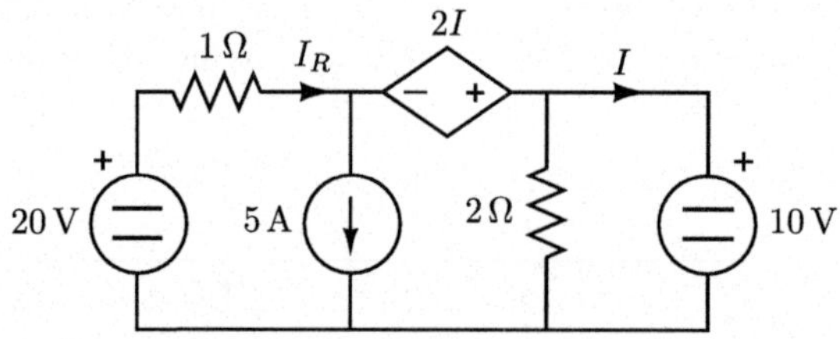

Fig. 2.26 Circuit considered in the problem under study

Solution The problem resolution will be approached using certain equivalents that allow simplifying the original circuit. In this sense, first, it can be observed that the 2 Ω resistor is connected in parallel with the 10 V voltage source and, consequently, the current flowing through it is 5 A. In the circuit of Fig. 2.27, this current has been represented as well as the current flowing through the voltage-dependent source. Also, node A has been presented which will be useful later on.

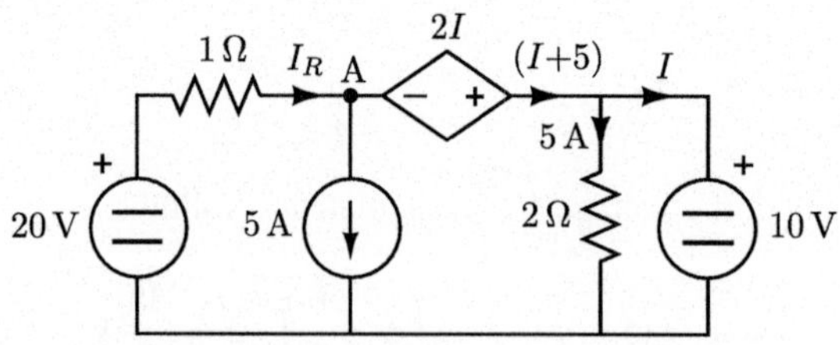

Fig. 2.27 Transformation of the original circuit

According to Fig. 2.27, the 10 V voltage source in parallel with the 2 Ω resistor can be replaced by its equivalent in which only one voltage source remains, as shown in Fig. 2.28. Likewise, the actual voltage source formed by the 20 V voltage source and the 1 Ω resistor has been replaced by its equivalent current source (Fig. 2.28).

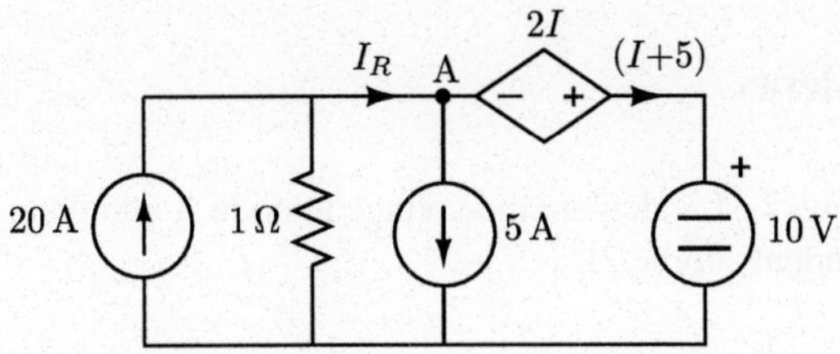

Fig. 2.28 Transformation of the original circuit

Next, the two current sources in parallel will be replaced by a single equivalent current source, as shown in Fig. 2.29.

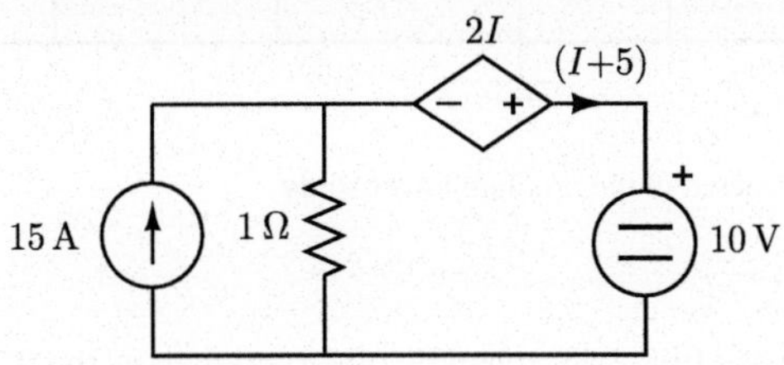

Fig. 2.29 Transformation of the original circuit

Finally, the actual current source formed by the 15 A current source and the 1 Ω resistor will be replaced by its equivalent voltage source as shown in Fig. 2.30. According to Fig. 2.30, applying the Kirchhoff voltage law results in:

$$10 - 15 + 1 \cdot (5 + I) - 2I = 0 \;\Rightarrow\; I = 0$$

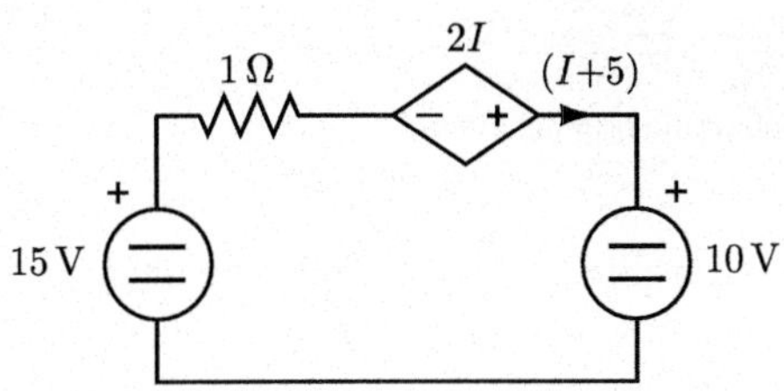

Fig. 2.30 Transformation of the original circuit

Once the current I is obtained, the current I_R can be calculated by applying Kirchhoff's current law at node A of the circuit in Fig. 2.27:

$$I_R = 5 + I + 5 = 10\,\text{A}.$$

2.12 Equivalents

In the circuit of Fig. 2.31, calculate the voltage at node A and the power supplied by the voltage-dependent source $2U_A$.

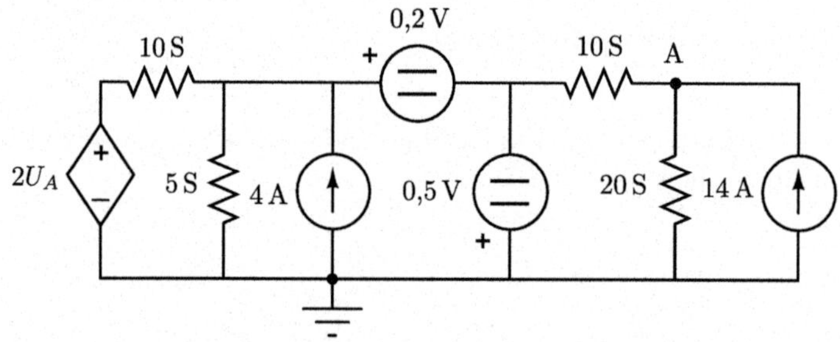

Fig. 2.31 Circuit considered in the problem under study

Solution According to Fig. 2.31, it can be observed that, as the 0, 5 V voltage source is connected, the circuit can be divided into two circuits, as shown in Fig. 2.32.

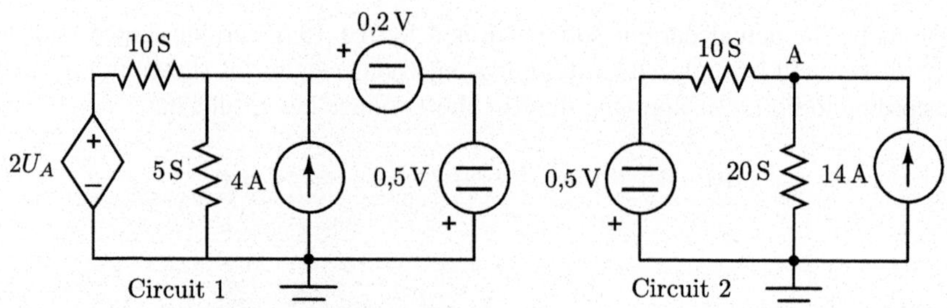

Fig. 2.32 Transformation of the original circuit

Considering circuit 2 of Fig. 2.32, after replacing the actual voltage source with its equivalent actual current source, applying the nodal analysis method results in:

$$(10 + 20) \cdot U_A = -0{,}5 \cdot 10 + 14 \;\Rightarrow\; U_A = 0{,}3\,\text{V}$$

On the other hand, circuit 1 of Fig. 2.32 can be simplified as shown in Fig. 2.33, where the 5 S conductance, the 4 A current source, and the 0,3 V voltage source have been replaced by an equivalent voltage source of 3 V.

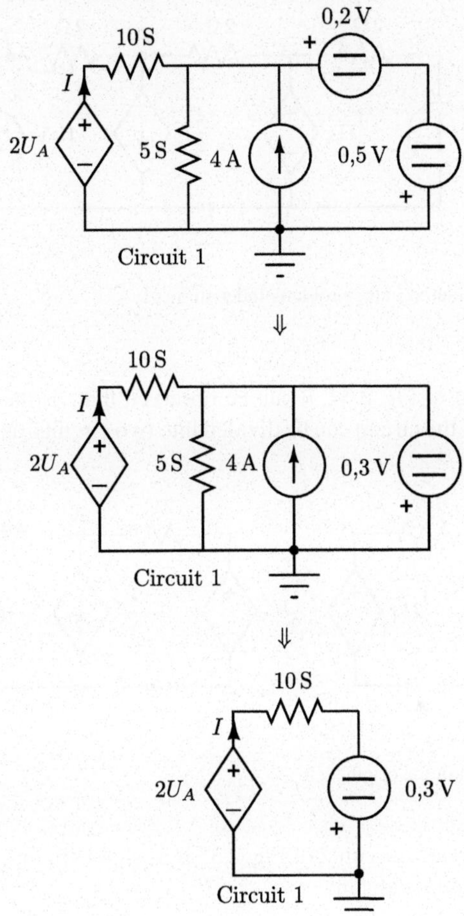

Fig. 2.33 Simplification of the original circuit 1

According to Fig. 2.33:

$$I = (2U_A + 0{,}3) \cdot 10 = (2 \cdot 0{,}3 + 0{,}3) \cdot 10 = 9\,\mathrm{A}$$

Therefore, the power supplied by the voltage-dependent source $2U_A$ is as follows:

$$P = 2U_A \cdot I = 2 \cdot 0{,}3 \cdot 9 = 5{,}4\,\mathrm{W}.$$

2.13 Equivalents

Determine the power generated by the 10 V source in the circuit of Fig. 2.34.

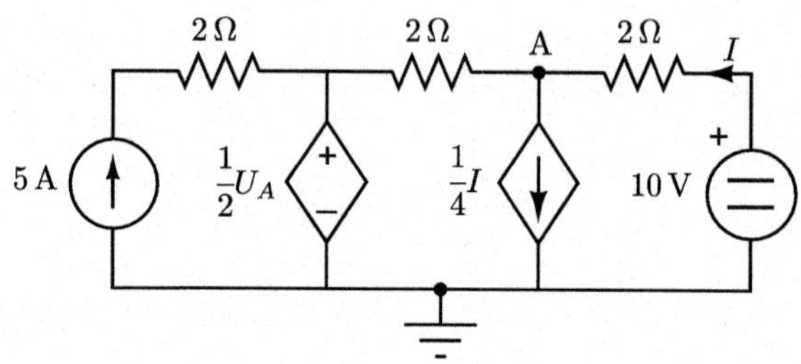

Fig. 2.34 Circuit considered in the problem under study

Solution According to Fig. 2.34, it can be observed that, as the voltage-dependent source is connected, the circuit can be divided into two circuits, as shown in Fig. 2.35.

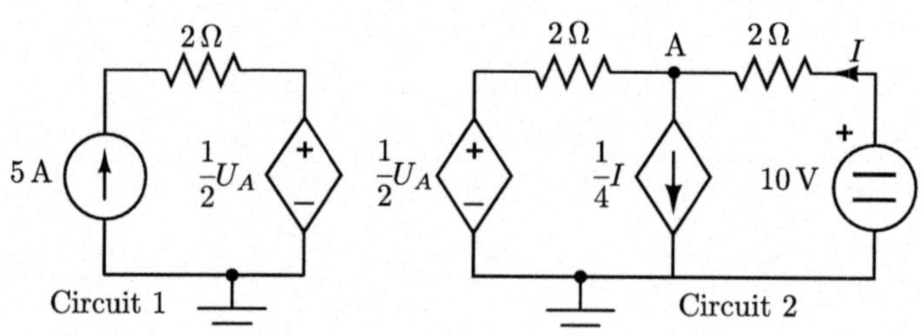

Fig. 2.35 Modification in the original circuit

By source conversion, circuit 2 of Fig. 2.35 is as shown in Fig. 2.36. Applying the nodal analysis method results in:

$$\left(\frac{1}{2}+\frac{1}{2}\right)\cdot U_A = \frac{1}{4}U_A - \frac{1}{4}I + 5$$

On the other hand:

$$I = 5 - \frac{U_A}{2} \Rightarrow U_A = 10 - 2\cdot I$$

Consequently:

$$\left(\frac{1}{2}+\frac{1}{2}\right)\cdot (10-2\cdot I) = \frac{1}{4}\cdot(10-2\cdot I) - \frac{1}{4}I + 5 \Rightarrow I = 2\,\text{A}$$

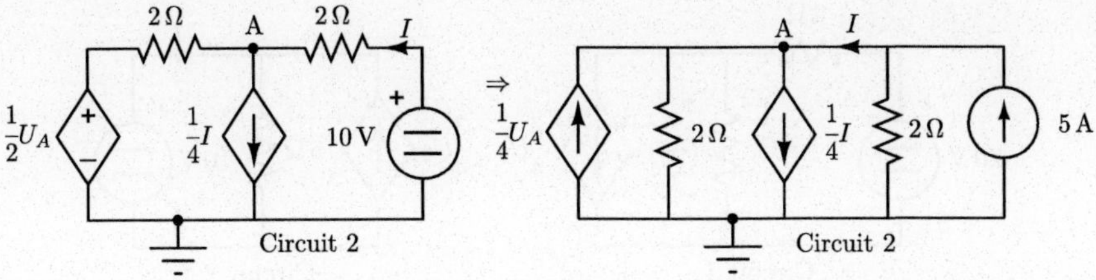

Fig. 2.36 Modification in the circuit 2

Therefore, the power generated by the 10 V voltage source is as follows:

$$P = 10 \cdot I = 10 \cdot 2 = 20\,\text{W}.$$

2.14 Equivalents

In the circuit of Fig. 2.37, determine the power generated by the 10 V source.

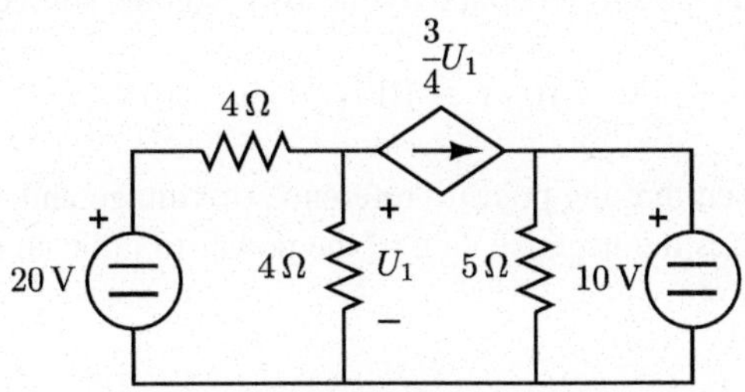

Fig. 2.37 Circuit considered in the problem under study

Solution According to Fig. 2.37, it can be observed that, as the current-dependent source is connected, the circuit can be divided into two circuits, as shown in Fig. 2.38. By replacing the actual voltage source with its equivalent actual current source, circuit 1 of Fig. 2.38 is as shown in Fig. 2.39.

According to Fig. 2.39, applying the nodal analysis method results in:

$$\left(\frac{1}{4} + \frac{1}{4}\right) \cdot U_1 = 5 - \frac{3}{4}U_1 \;\Rightarrow\; U_1 = 4\,\text{V}$$

On the other hand, according to circuit 2 of Fig. 2.38, applying Kirchhoff's current law results in:

$$I = \frac{10}{5} - \frac{3}{4}U_1 = \frac{10}{5} - \frac{3}{4} \cdot 4 = -1\,\text{A}$$

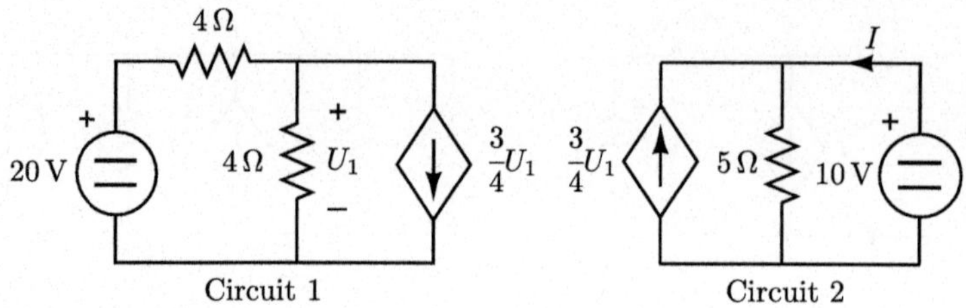

Fig. 2.38 Modification in the original circuit

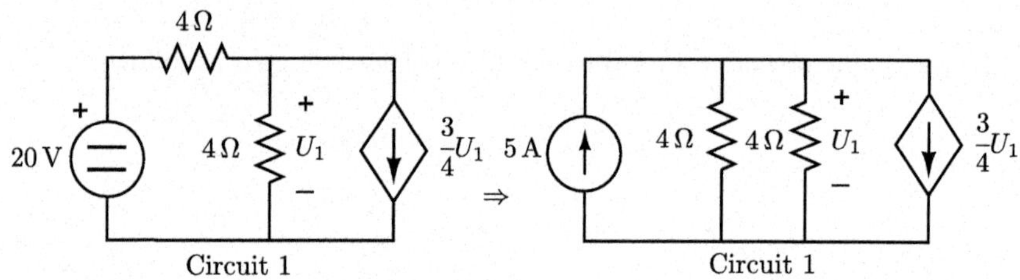

Fig. 2.39 Modification in the original circuit

Consequently, the power generated by the 10 V voltage source is as follows:

$$P = 10 \cdot I = 10 \cdot (-1) = -10\,\text{W}$$

It can be observed that the polarity reference of voltage and current in the 10 V voltage source causes the negative sign of the power to indicate that it is absorbing 10 W.

2.15 Thévenin and Norton Equivalents

In the circuit of Fig. 2.40, find the Thévenin and Norton equivalents between terminals a and b.

Solution First, the circuit will be simplified by connecting the 6 Ω and 12 Ω resistors in parallel and replacing the actual current source with its equivalent actual voltage source, resulting in the circuit shown in Fig. 2.41. It can be seen that the current I has been introduced to facilitate the calculation of the open-circuit voltage between terminals a and b, the Thévenin voltage.

According to Fig. 2.41, the current I is calculated as follows:

$$I = \frac{24 - 12}{4 + 12} = \frac{3}{4}\,\text{A}$$

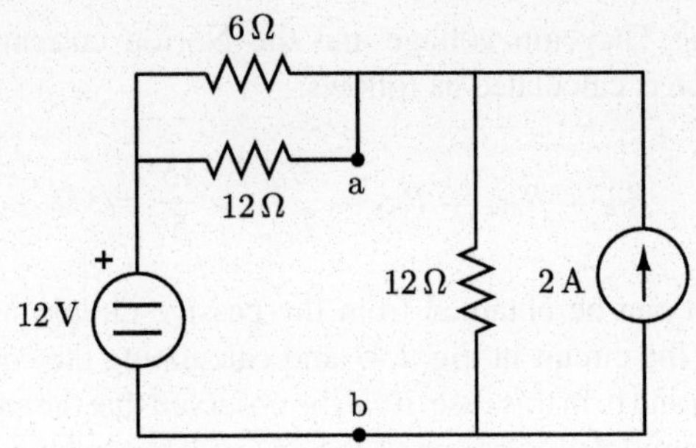

Fig. 2.40 Circuit considered in the problem under study

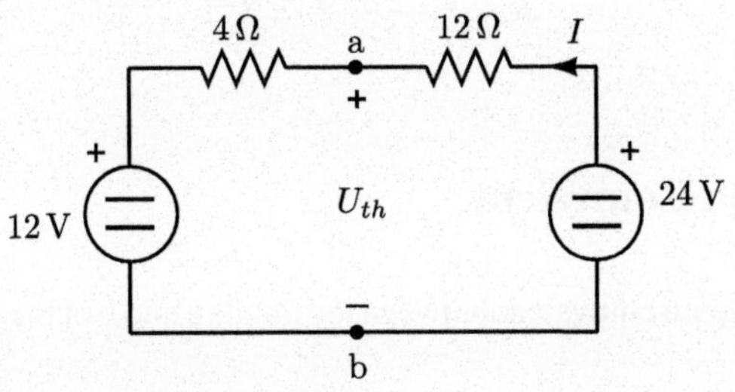

Fig. 2.41 Modification in the original circuit

Once the current I is obtained, the Thévenin voltage is as follows:

$$U_{th} = 24 - 12 \cdot I = 24 - 12 \cdot \frac{3}{4} = 15\,\mathrm{V}$$

Next, the short-circuit current, the Norton current, is calculated, for which the circuit of Fig. 2.42 is considered, according to which:

$$I_{nor} = \frac{12}{4} + \frac{24}{12} = 3 + 2 = 5\,\mathrm{A}$$

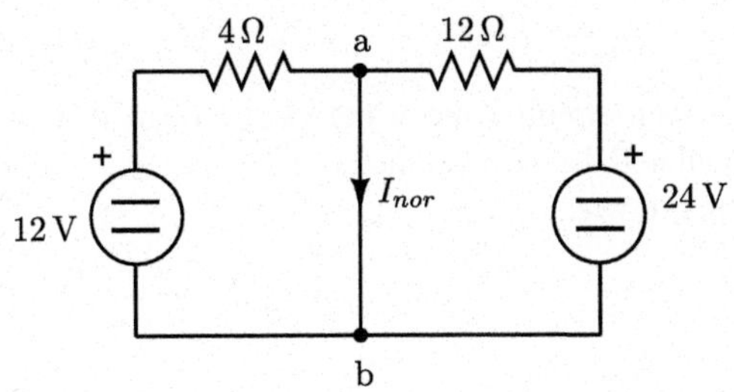

Fig. 2.42 Circuit with terminal a and b short-circuited

Finally, once the Thévenin voltage and the Norton current are obtained, the equivalent resistance is calculated as follows:

$$R_{th} = R_{nor} = R_{eq} = \frac{U_{th}}{I_{nor}} = \frac{15}{5} = 3\,\Omega$$

This same result can be obtained from the passive circuit, nullifying the independent sources of the circuit in Fig. 2.41 and calculating the equivalent resistance between terminals a and b. In this case, it can be observed that the equivalent resistance is obtained by the parallel association of the 4 Ω resistor and the 12 Ω resistor:

$$R_{eq} = \frac{12 \cdot 4}{12 + 4} = 3\,\Omega.$$

2.16 Thévenin Equivalent

Calculate the Thévenin equivalent between terminals a and b of the circuit in Fig. 2.43.

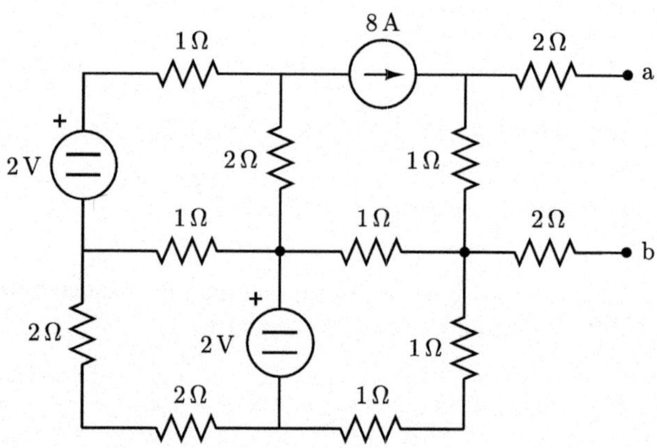

Fig. 2.43 Circuit considered in the problem under study

Solution First, the open-circuit voltage between terminals a and b, the Thévenin voltage, will be calculated. Looking at the circuit in Fig. 2.44, it is easy to verify that the voltage U_{th} is as follows:

$$U_{th} = 8 \cdot 1 = 8\,\text{V}$$

Next, the current when terminals a and b are short-circuited, the Norton current, will be calculated, as shown in Fig. 2.45.

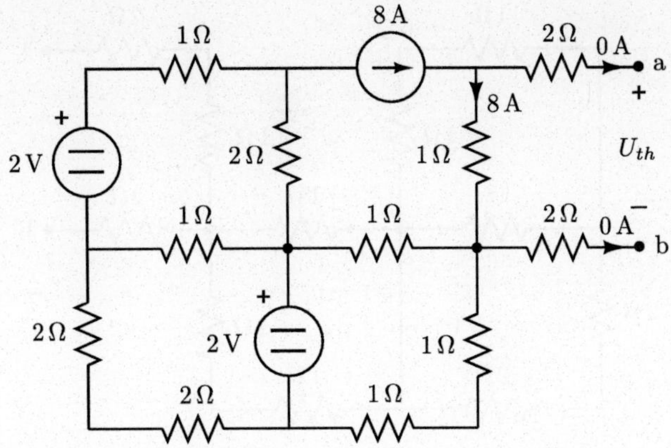

Fig. 2.44 Modification in the original circuit

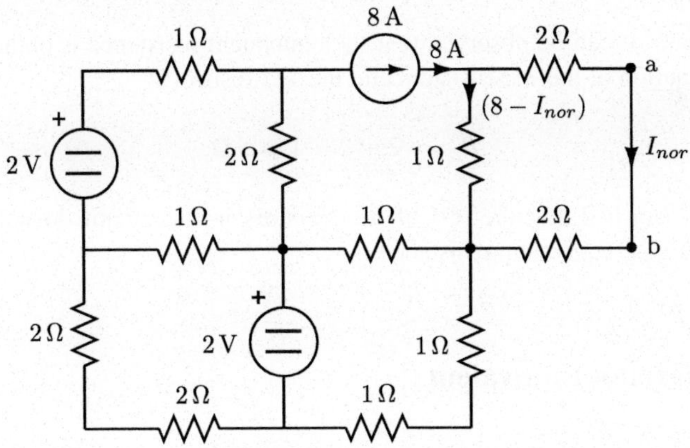

Fig. 2.45 Modification in the original circuit

According to Fig. 2.45, applying Kirchhoff's voltage law to the loop formed by the 1 Ω resistor and the two 2 Ω resistors results in:

$$1 \cdot (8 - I_{nor}) - 2 \cdot I_{nor} - 2 \cdot I_{nor} = 0 \ \Rightarrow \ I_{nor} = \frac{8}{5}\,\text{A}$$

Once the Thévenin voltage and the Norton current are obtained, the equivalent resistance between terminals a and b is calculated as follows:

$$R_{eq} = \frac{U_{th}}{I_{nor}} = \frac{8}{8/5} = 5\,\Omega$$

The calculation of the equivalent resistance could also have been done from the passive circuit in Fig. 2.43, as shown in Fig. 2.46.

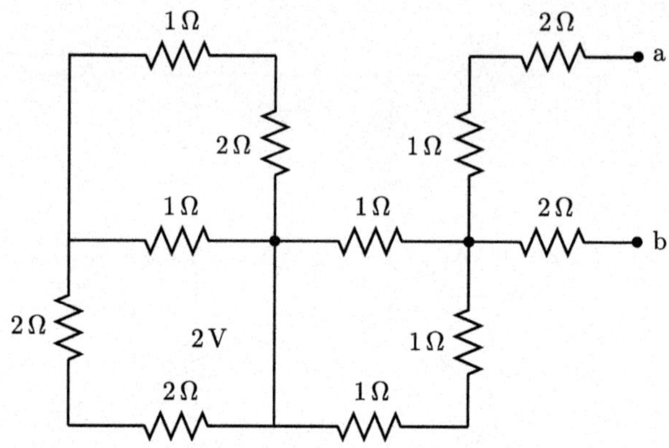

Fig. 2.46 Passive circuit for calculation of equivalent resistance

In this case, it can be observed that the equivalent resistance is obtained by the series association of the 2 Ω resistors and the 1 Ω resistor:

$$R_{eq} = 2 + 2 + 1 = 5\,\Omega$$

It can be verified that the rest of the resistors in the circuit do not affect the calculation of the equivalent resistance.

2.17 Thévenin Equivalent

In the circuit of Fig. 2.47, calculate the Thévenin equivalent between terminals a and b. Using the obtained equivalent, calculate the power absorbed by a 5 Ω resistor connected between terminals a and b.

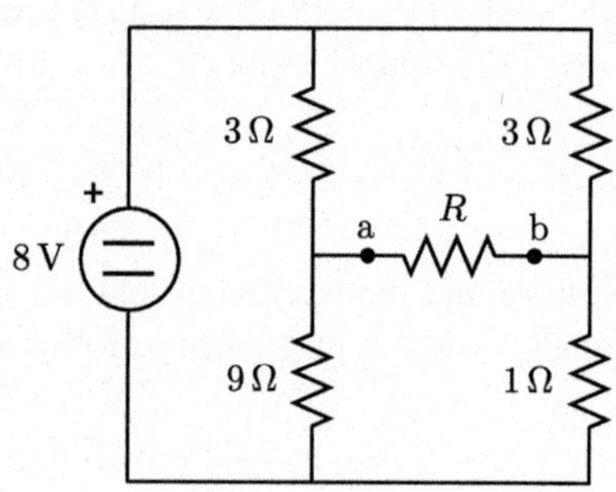

Fig. 2.47 Circuit considered in the problem under study

Solution First, the open-circuit voltage between terminals a and b, the Thévenin voltage, will be calculated. To do this, it is necessary to remove resistor R, as it is not part of the equivalent. The resulting circuit is shown in Fig. 2.48.

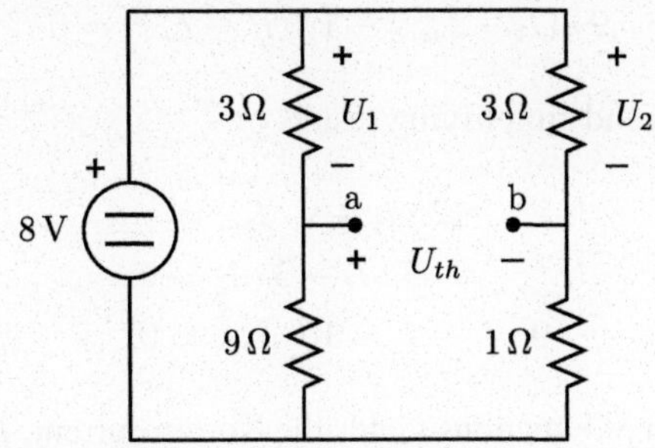

Fig. 2.48 Circuit without resistor R

According to Fig. 2.48, applying the voltage divider concept yields the voltages U_1 and U_2:

$$U_1 = 8 \cdot \frac{3}{3+9} = 2\,\text{V}$$

$$U_2 = 8 \cdot \frac{3}{3+1} = 6\,\text{V}$$

Once U_1 and U_2 are calculated, the Thévenin voltage can be calculated by applying Kirchhoff's voltage law:

$$U_1 + U_{th} - U_2 = 0 \;\Rightarrow\; U_{th} = U_2 - U_1 = 6 - 2 = 4\,\text{V}$$

Next, the Norton current is calculated by short-circuiting terminals a and b, as shown in Fig. 2.49. The currents I_1 and I_2 are indicated to facilitate the resolution. Likewise, Kirchhoff's current law is applied at the nodes near terminals a and b.

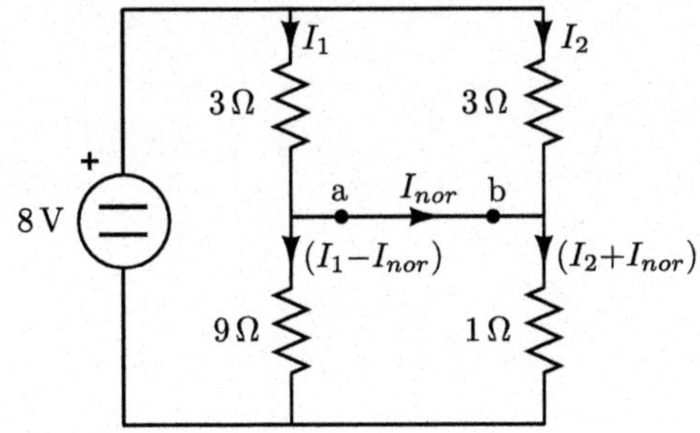

Fig. 2.49 Circuit with terminals a and b short-circuited

According to Fig. 2.49, applying Kirchhoff's voltage law results in:

$$8 - 9 \cdot (I_1 - I_{nor}) - 3 \cdot I_1 = 0$$
$$3 \cdot I_1 - 3 \cdot I_2 = 0$$
$$9 \cdot (I_1 - I_{nor}) - 1 \cdot (I_2 + I_{nor}) = 0$$

Rearranging terms and simplifying results in:

$$-12 \cdot I_1 + 9 \cdot I_{nor} = -8$$
$$3 \cdot I_1 - 3 \cdot I_2 = 0$$
$$9 \cdot I_1 - I_2 - 10 \cdot I_{nor} = 0$$

Solving the system of equations yields the Norton current, I_{nor}:

$$I_{nor} = \frac{4}{3}\,\text{A}$$

Once the Thévenin voltage and the Norton current are obtained, the equivalent resistance between terminals a and b, Thévenin resistance, is calculated as follows:

$$R_{th} = \frac{U_{th}}{I_{nor}} = \frac{4}{4/3} = 3\,\Omega$$

The equivalent resistance could also have been obtained from the passive circuit shown in Fig. 2.47.

According to Fig. 2.50, the equivalent resistance corresponds to the parallel connection of the 2 Ω resistor and the two 4 Ω resistors:

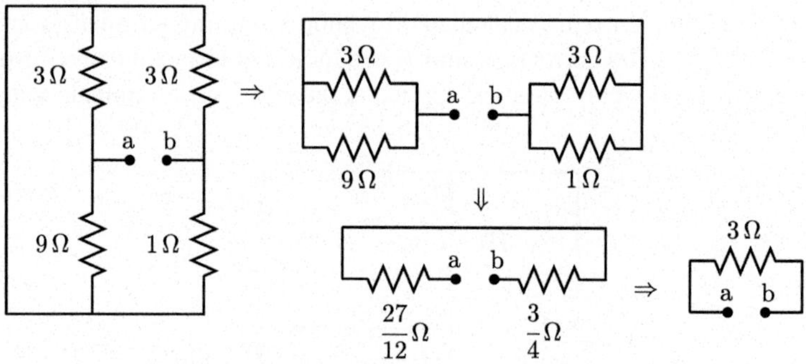

Fig. 2.50 Passive circuit

$$R_{eq} = 2//4//4 = \frac{1}{\frac{1}{2}+\frac{1}{4}+\frac{1}{4}} = 1\,\Omega$$

Now that the Thévenin equivalent between terminals a and b has been obtained, a 5 Ω resistor will be connected between these terminals. The resulting circuit is shown in Fig. 2.51.

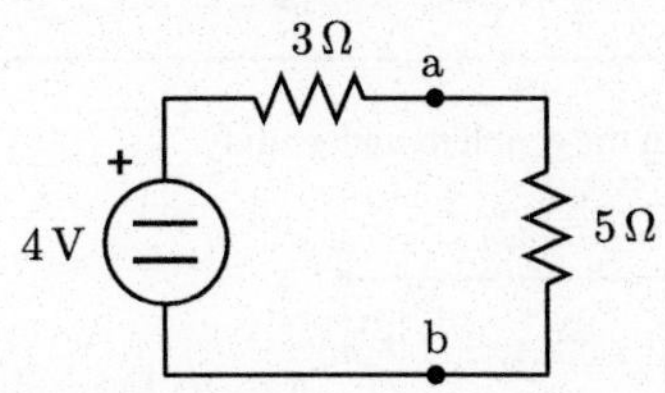

Fig. 2.51 Circuit with a 5 Ω resistor connected

According to Fig. 2.51:

$$P_{5\Omega} = \frac{U_{ab}^2}{5}$$

The voltage U_{ab} can be calculated by applying the voltage divider concept:

$$U_{ab} = 4 \cdot \frac{5}{5+3} = \frac{5}{2}\,\mathrm{V}$$

Therefore, the power absorbed by the 5 Ω resistor is:

$$P_{5\Omega} = \frac{U_{ab}^2}{5} = \frac{(5/2)^2}{5} = \frac{5}{4}\,\mathrm{W}.$$

2.18 Thévenin and Norton Equivalents

In the circuit of Fig. 2.52, find the Thévenin and Norton equivalents between terminals a and b.

Solution First, the open-circuit voltage between terminals a and b, the Thévenin voltage, will be calculated. To do this, the circuit shown in Fig. 2.53 will be solved, where the parallel connection of the 6 V voltage source and the 4 Ω resistor has been replaced by the voltage source alone. The resolution will be done using the mesh analysis method.

According to Fig. 2.53:

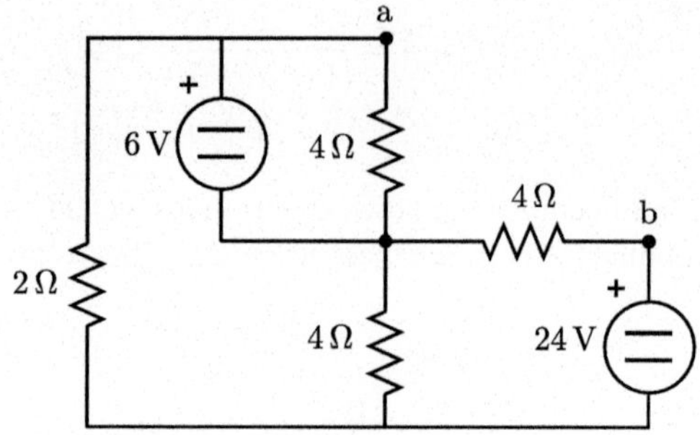

Fig. 2.52 Circuit considered in the problem under study

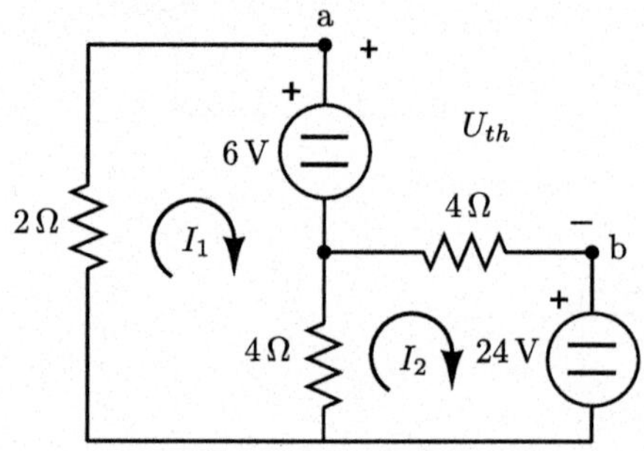

Fig. 2.53 Circuit with the 6 V voltage source and the 4 Ω resistor replaced

$$\begin{pmatrix} 2+4 & -4 \\ -4 & 4+4 \end{pmatrix} \begin{pmatrix} I_1 \\ I_2 \end{pmatrix} = \begin{pmatrix} -6 \\ -24 \end{pmatrix}$$

Solving yields the mesh currents:

$$I_1 = \frac{-9}{2}\,\mathrm{A}\ ;\ I_2 = \frac{-21}{4}\,\mathrm{A}$$

Once the mesh currents are obtained, the voltage U_{th} is found by applying Kirchhoff's voltage law:

$$U_{th} - 4 \cdot I_2 - 6 = 0 \;\Rightarrow\; U_{th} = 6 + 4 \cdot I_2 = 6 + 4 \cdot \frac{-21}{4} = 6 - 21 = -15\,\mathrm{V}$$

Next, the Norton current is calculated by short-circuiting terminals a and b, as shown in Fig. 2.54. The currents I_1, I_2, and I_3 are indicated to apply the mesh analysis method.

According to Fig. 2.54:

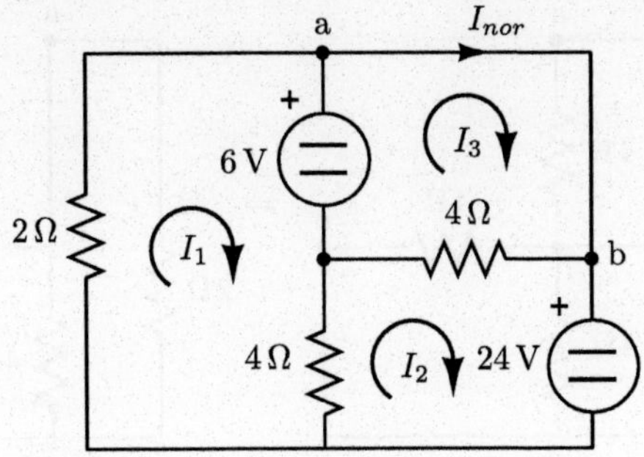

Fig. 2.54 Circuit with terminals a and b short-circuited

$$\begin{pmatrix} 2+4 & -4 & 0 \\ -4 & 4+4 & -4 \\ 0 & -4 & 4 \end{pmatrix} \begin{pmatrix} I_1 \\ I_2 \\ I_3 \end{pmatrix} = \begin{pmatrix} -6 \\ -24 \\ 6 \end{pmatrix}$$

Solving yields the mesh currents:

$$I_1 = -12\,\text{A} \ ; \ I_2 = -16,5\,\text{A} \ ; \ I_3 = -15\,\text{A}$$

Therefore, the Norton current is:

$$I_{nor} = I_3 = -15\,\text{A}$$

Once the Thévenin voltage and the Norton current are obtained, the equivalent resistance between terminals a and b is calculated as follows:

$$R_{th} = \frac{U_{th}}{I_{nor}} = \frac{-15}{-15} = 1\,\Omega$$

The equivalent resistance could also have been obtained from the passive circuit shown in Fig. 2.55.

According to Fig. 2.55, the equivalent resistance is the parallel connection of the 2 Ω resistor and the two 4 Ω resistors:

$$R_{eq} = 2//4//4 = \frac{1}{\frac{1}{2} + \frac{1}{4} + \frac{1}{4}} = 1\,\Omega.$$

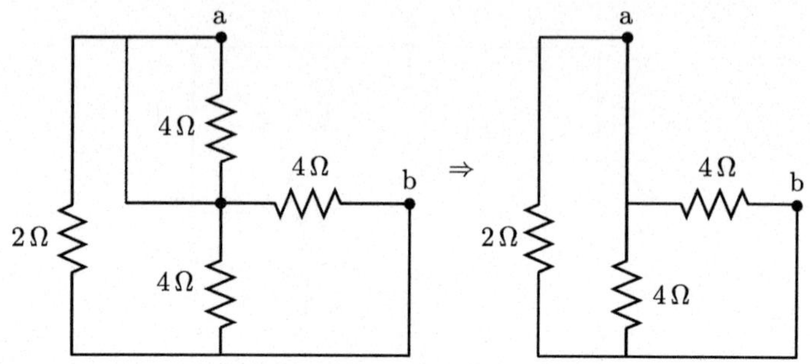

Fig. 2.55 Passive circuit for problem under study

2.19 Thévenin and Norton Equivalents

Determine the Thévenin and Norton equivalents between terminals a and b of the circuit in Fig. 2.56.

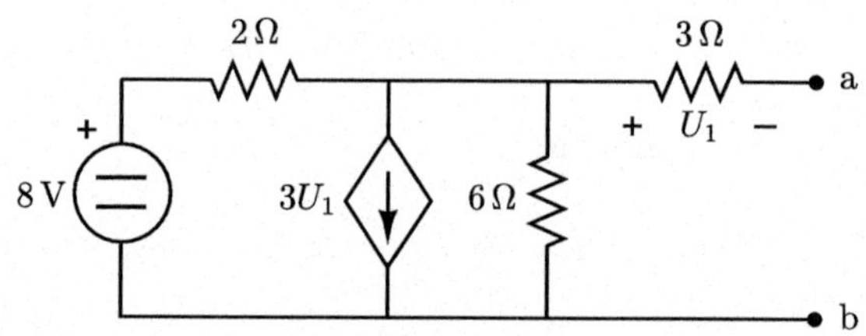

Fig. 2.56 Circuit considered in the problem under study

Solution First, the open-circuit voltage between terminals a and b, Thévenin voltage, will be calculated. To do this, solve the circuit shown in Fig. 2.57, where it can be observed that $U_1 = 0\,\text{V}$. Therefore, applying the voltage divider concept:

$$U_{th} = 8 \cdot \frac{6}{2+6} = 6\,\text{V}$$

Next, calculate the Norton current, obtained by short-circuiting terminals a and b, as shown in Fig. 2.58.

According to Fig. 2.58, applying the nodal analysis, after replacing the real voltage source with its equivalent real current source:

$$\left(\frac{1}{2} + \frac{1}{6} + \frac{1}{3}\right) \cdot U_1 = \frac{8}{2} - 3U_1 \quad \Rightarrow \quad U_1 = 1\,\text{V}$$

From voltage U_1, calculate the Norton current as follows:

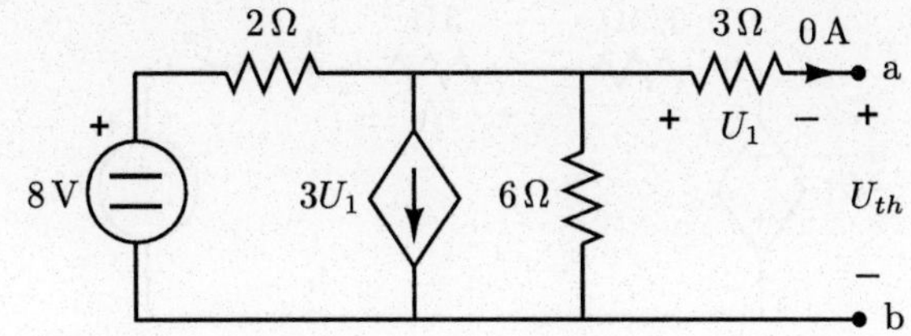

Fig. 2.57 Circuit considered in the problem under study

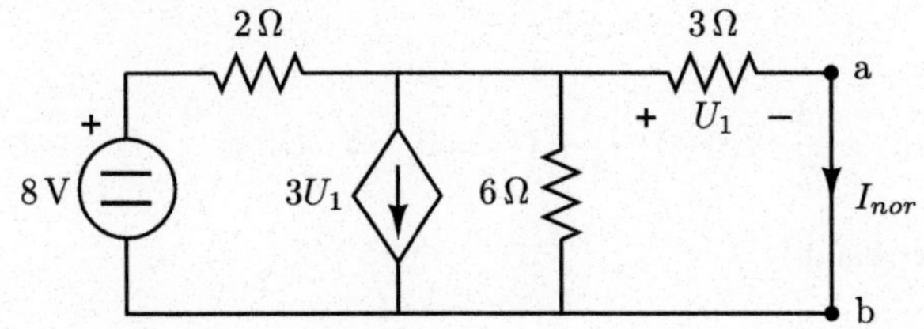

Fig. 2.58 Circuit with terminals a and b short-circuited

$$I_{nor} = \frac{U_1}{3} = \frac{1}{3}\,\text{A}$$

Once Thévenin voltage and Norton current are obtained, the equivalent resistance between terminals a and b is calculated as follows:

$$R_{eq} = \frac{U_{th}}{I_{nor}} = \frac{6}{1/3} = 18\,\Omega$$

In this case, as the circuit has a dependent source, the equivalent resistance cannot be obtained by associating resistors in the passive circuit. Instead, a test source connected between terminals a and b of the passive circuit could be used as shown in Fig. 2.59.

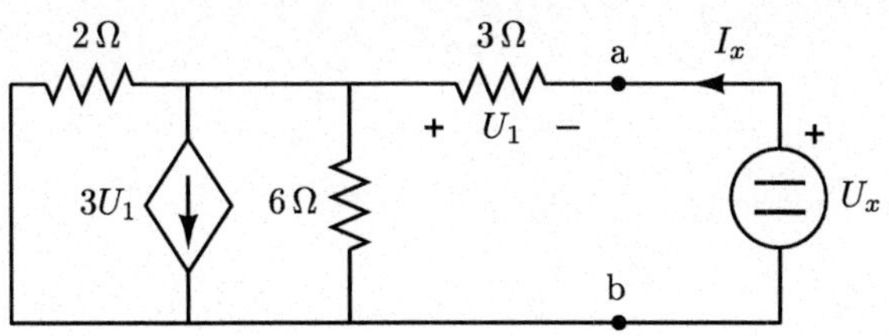

Fig. 2.59 Circuit with a test voltage source

The circuit in Fig. 2.59 can be simplified by associating the resistors of 2 Ω and 6 Ω and replacing the real current-dependent source with its equivalent real voltage-dependent source as shown in Fig. 2.60. Applying Kirchhoff's voltage law:

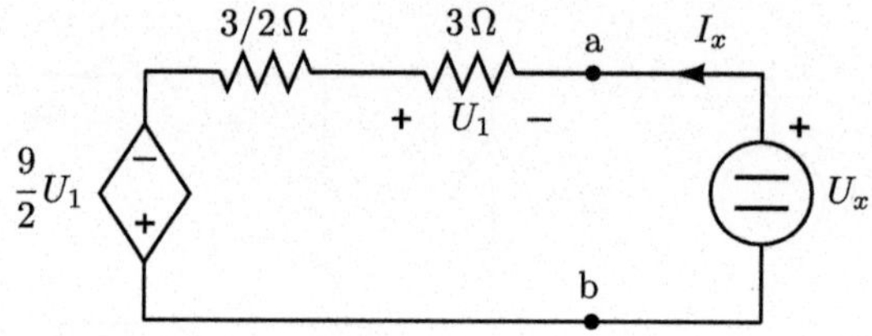

Fig. 2.60 Simplified circuit

$$\frac{9}{2}U_1 - \frac{3}{2} \cdot I_x - 3 \cdot I_x + U_x = 0 \Rightarrow U_x = \frac{9}{2} \cdot I_x - \frac{9}{2}U_1$$

On the other hand:

$$U_1 = -3 \cdot I_x$$

Consequently:

$$U_x = \frac{9}{2} \cdot I_x - \frac{9}{2}U_1 = \frac{9}{2} \cdot I_x - \frac{9}{2} \cdot (-3 \cdot I_x) = 18 \cdot I_x \Rightarrow R_{eq} = \frac{U_x}{I_x} = 18\,\Omega.$$

2.20 Thévenin and Norton Equivalents

Calculate the Thévenin and Norton equivalents between terminals a and b of the circuit in Fig. 2.61.

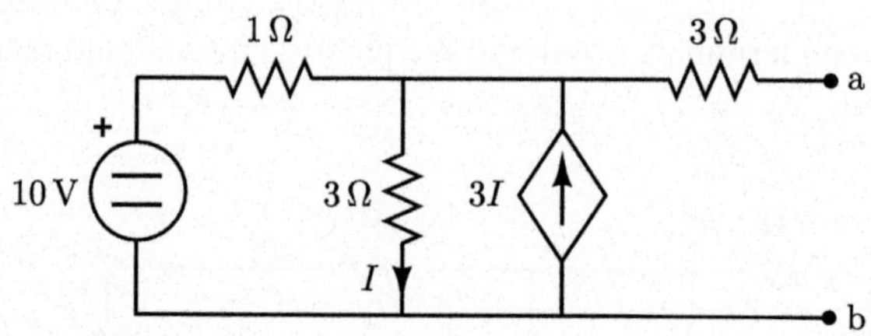

Fig. 2.61 Circuit considered in the problem under study

Solution First, the open-circuit voltage between terminals a and b, Thévenin voltage, will be calculated. To do this, solve the circuit shown in Fig. 2.62. Applying Kirchhoff's voltage law:

$$10 = -2 \cdot I + 3 \cdot I = I$$

On the other hand:

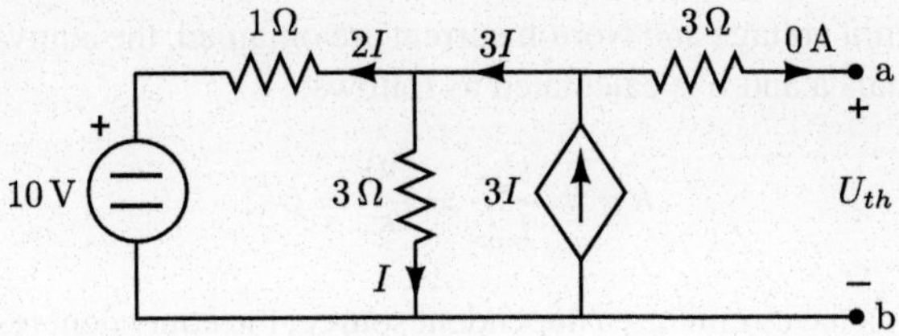

Fig. 2.62 Circuit considered in the problem under study

$$U_{th} = 3 \cdot I = 3 \cdot 10 = 30\,\text{V}$$

Next, calculate the Norton current, obtained by short-circuiting terminals a and b, as shown in Fig. 2.63. It can be observed that voltage U has been introduced into the circuit in order to apply nodal analysis.

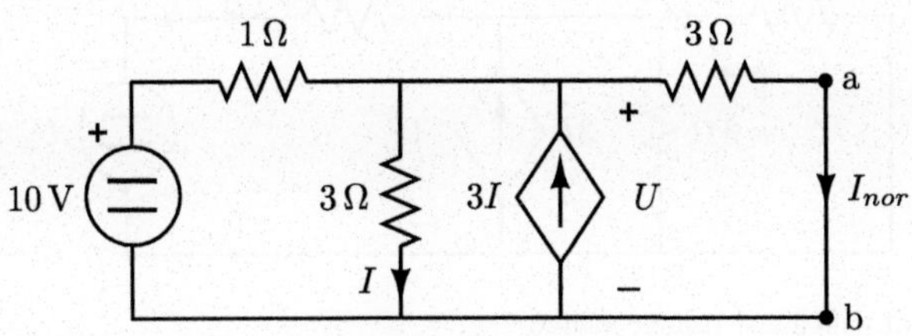

Fig. 2.63 Circuit with terminals a and b short-circuited

According to Fig. 2.63, after replacing the real voltage source with its equivalent real current source and applying nodal analysis:

$$\left(\frac{1}{1} + \frac{1}{3} + \frac{1}{3}\right) \cdot U = \frac{10}{1} + 3I \;\Rightarrow\; \frac{5}{3} \cdot U = 10 + 3I$$

On the other hand:

$$I = \frac{U}{3}$$

Consequently:

$$\frac{5}{3} \cdot U = 10 + 3 \cdot \frac{U}{3} \;\Rightarrow\; U = 15\,\text{V}$$

Once voltage U is obtained, the Norton current is as follows:

$$I_{nor} = \frac{U}{3} = \frac{15}{3} = 5\,\text{A}$$

Once Thévenin voltage and Norton current are obtained, the equivalent resistance between terminals a and b is calculated as follows:

$$R_{eq} = \frac{U_{th}}{I_{nor}} = \frac{30}{5} = 6\,\Omega$$

In this case, as the circuit has a dependent source, the equivalent resistance cannot be obtained by associating resistors in the passive circuit. Instead, a test source connected between terminals a and b of the passive circuit could be used as shown in Fig. 2.64. According to Fig. 2.64, applying nodal analysis:

$$\left(\frac{1}{1}+\frac{1}{3}+\frac{1}{3}\right)\cdot U = \frac{U_x}{3} + 3I \;\Rightarrow\; \frac{5}{3}\cdot U = \frac{U_x}{3} + 3I$$

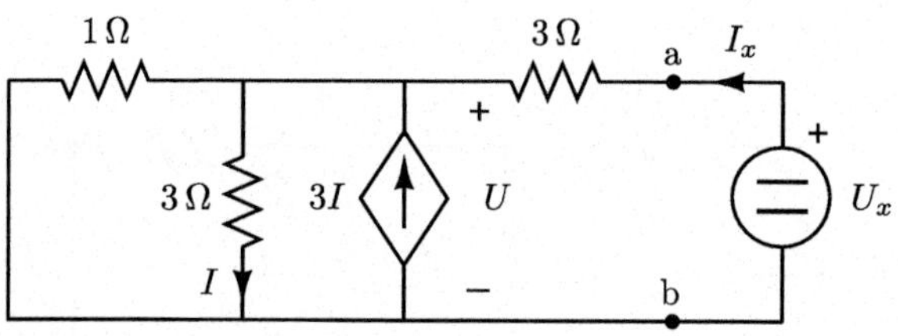

Fig. 2.64 Circuit with a test voltage source

On the other hand:

$$I = \frac{U}{3}$$

Consequently:

$$\frac{5}{3}\cdot U = \frac{U_x}{3} + 3\cdot\frac{U}{3} \;\Rightarrow\; U = \frac{U_x}{2}$$

According to Fig. 2.64, applying Kirchhoff's voltage law results in the following relationship:

$$U_x = 3\cdot I_x + U$$

Therefore:

$$U_x = 3\cdot I_x + \frac{U_x}{2} \;\Rightarrow\; R_{eq} = \frac{U_x}{I_x} = 6\,\Omega.$$

2.21 Thévenin and Norton Equivalents

Calculate the Thévenin and Norton equivalents between terminals a and b of the circuit in Fig. 2.65.

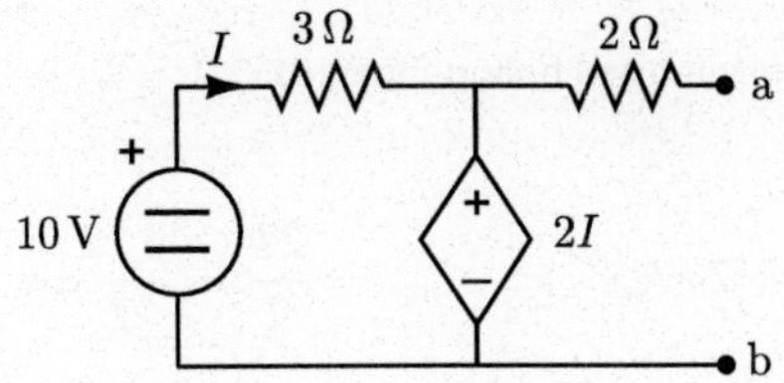

Fig. 2.65 Circuit considered in the problem under study

Solution First, the open circuit voltage between terminals a and b, Thévenin voltage, will be calculated. To do this, the circuit shown in Fig. 2.66 will be solved.

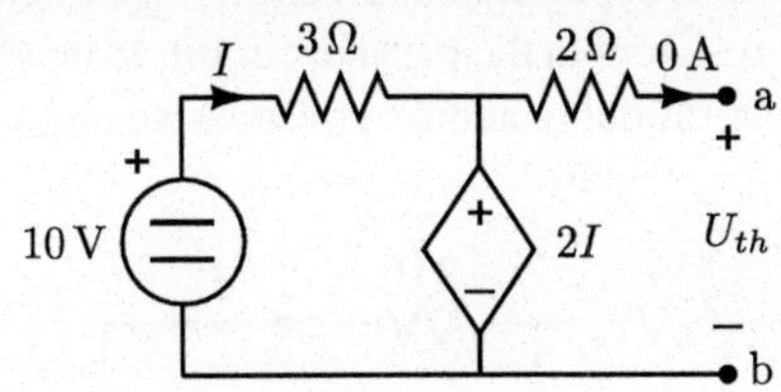

Fig. 2.66 Circuit considered in the problem under study

According to Fig. 2.66:

$$10 = 3 \cdot I + 2I = 5I \;\Rightarrow\; I = \frac{10}{5} = 2\,\text{A}$$

Therefore:

$$U_{th} = 2I = 2 \cdot 2 = 4\,\text{V}$$

Next, the Norton current is calculated, which is obtained by short-circuiting terminals a and b, as shown in Fig. 2.67.

According to Fig. 2.67:

$$10 = 3 \cdot I + 2I = 5I \;\Rightarrow\; I = \frac{10}{5} = 2\,\text{A}$$

On the other hand:

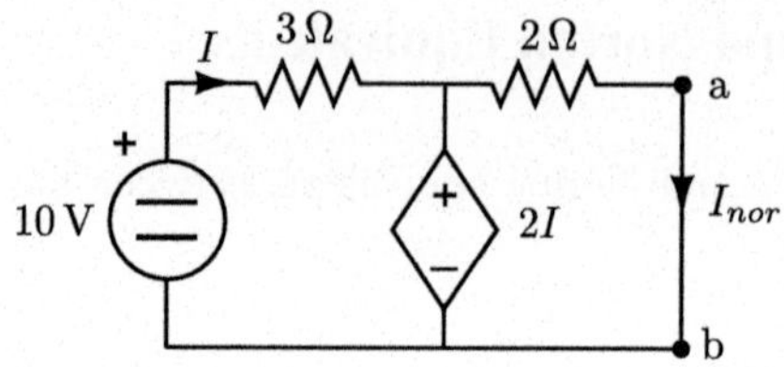

Fig. 2.67 Circuit with terminals a and b short-circuited

$$2I = 2 \cdot I_{nor} \;\Rightarrow\; I_{nor} = \frac{2I}{2} = \frac{2 \cdot 2}{2} = 2\,\text{A}$$

Once the Thévenin voltage and Norton current are calculated, the equivalent resistance between terminals a and b is obtained as follows:

$$R_{eq} = \frac{U_{th}}{I_{nor}} = \frac{4}{2} = 2\,\Omega$$

Next, an alternative way to calculate the equivalent resistance is shown. In this case, since the circuit has a dependent source, the equivalent resistance cannot be obtained by associating resistors in the passive circuit. Instead, a test source could be used connected between terminals a and b of the passive circuit as shown in Fig. 2.68.

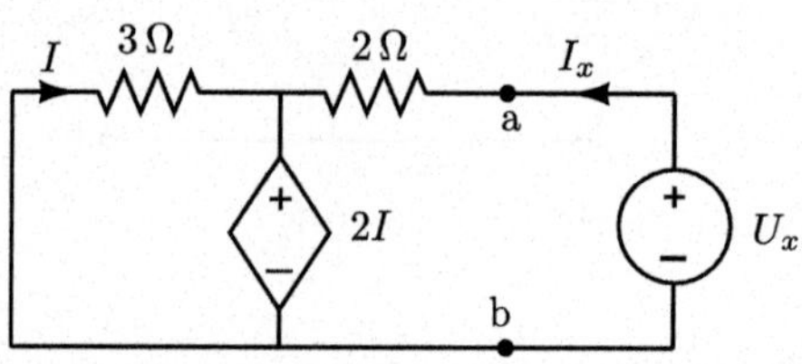

Fig. 2.68 Circuit with a test voltage source

According to Fig. 2.68:

$$U_x = 2 \cdot I_x + 2I$$

On the other hand:

$$2I + 3 \cdot I = 0 \;\Rightarrow\; I = 0\,\text{A}$$

Consequently:

$$U_x = 2 \cdot I_x + 2I = 2 \cdot I_x \;\Rightarrow\; R_{eq} = \frac{U_x}{I_x} = 2\,\Omega.$$

2.22 Thévenin and Norton Equivalents

Determine the Thévenin and Norton equivalents, from terminals a and b, of the circuit in Fig. 2.69.

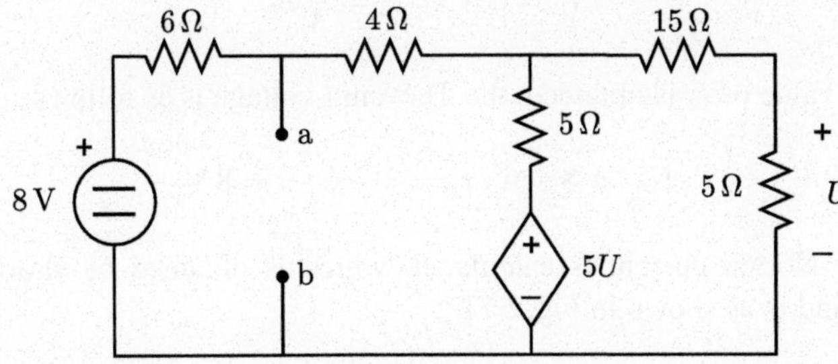

Fig. 2.69 Circuit considered in the problem under study

Solution First, the open circuit voltage between terminals a and b, Thévenin voltage, will be calculated. To do this, the circuit shown in Fig. 2.70 will be solved.

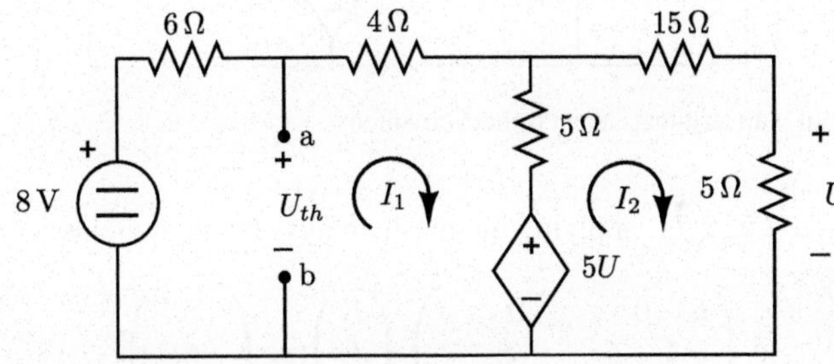

Fig. 2.70 Circuit considered in the problem under study

According to Fig. 2.70, applying the mesh analysis method yields:

$$\begin{pmatrix} 6+4+5 & -5 \\ -5 & 5+15+5 \end{pmatrix} \begin{pmatrix} I_1 \\ I_2 \end{pmatrix} = \begin{pmatrix} 8-5U \\ 5U \end{pmatrix}$$

On the other hand:

$$U = 5 \cdot I_2$$

Therefore:

$$\begin{pmatrix} 15 & -5 \\ -5 & 25 \end{pmatrix} \begin{pmatrix} I_1 \\ I_2 \end{pmatrix} = \begin{pmatrix} 8-5U \\ 5U \end{pmatrix} = \begin{pmatrix} 8-5\cdot(5\cdot I_2) \\ 5\cdot(5\cdot I_2) \end{pmatrix} = \begin{pmatrix} 8-25\cdot I_2 \\ 25\cdot I_2 \end{pmatrix}$$

Rearranging terms:

$$\begin{pmatrix} 15 & 20 \\ -5 & 0 \end{pmatrix} \begin{pmatrix} I_1 \\ I_2 \end{pmatrix} = \begin{pmatrix} 8 \\ 0 \end{pmatrix}$$

and solving

$$I_1 = 0\,\text{A} \; ; \; I_2 = \frac{2}{5}\,\text{A}$$

Once the value of I_1 is obtained, the Thévenin voltage is as follows:

$$U_{th} = 8 - 6 \cdot I_1 = 8 - 6 \cdot 0 = 8\,\text{V}$$

Next, the Norton current is calculated, which is obtained by short-circuiting terminals a and b, as shown in Fig. 2.71.

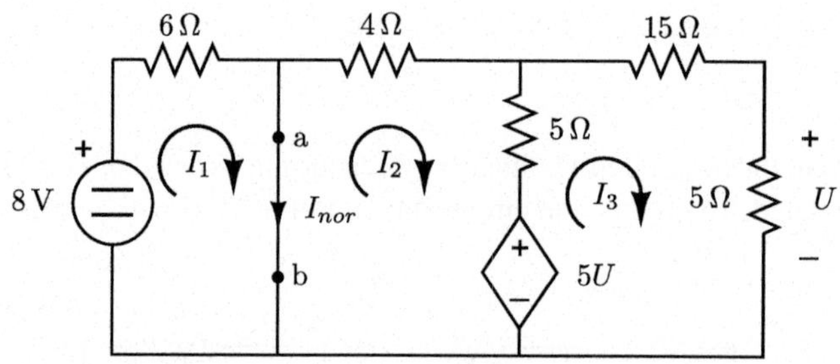

Fig. 2.71 Circuit with terminals a and b short-circuited

According to Fig. 2.71, applying the mesh analysis method yields:

$$\begin{pmatrix} 6 & 0 & 0 \\ 0 & 4+5 & -5 \\ 0 & -5 & 5+15+5 \end{pmatrix} \begin{pmatrix} I_1 \\ I_2 \\ I_3 \end{pmatrix} = \begin{pmatrix} 8 \\ -5U \\ 5U \end{pmatrix}$$

On the other hand:

$$U = 5 \cdot I_3$$

Therefore:

$$\begin{pmatrix} 6 & 0 & 0 \\ 0 & 9 & -5 \\ 0 & -5 & 25 \end{pmatrix} \begin{pmatrix} I_1 \\ I_2 \\ I_3 \end{pmatrix} = \begin{pmatrix} 8 \\ -5 \cdot (5 \cdot I_3) \\ 5 \cdot (5 \cdot I_3) \end{pmatrix} = \begin{pmatrix} 8 \\ -25 \cdot I_3 \\ 25 \cdot I_3 \end{pmatrix}$$

Rearranging terms:

$$\begin{pmatrix} 6 & 0 & 0 \\ 0 & 9 & 20 \\ 0 & -5 & 0 \end{pmatrix} \begin{pmatrix} I_1 \\ I_2 \\ I_3 \end{pmatrix} = \begin{pmatrix} 8 \\ 0 \\ 0 \end{pmatrix}$$

and solving

$$I_1 = \frac{8}{6}\,\text{A} \ ; \ \ I_2 = 0\,\text{A} \ ; \ \ I_3 = 0\,\text{A}$$

Thus, the Norton current is as follows:

$$I_{nor} = I_1 - I_2 = \frac{8}{6}\,\text{A}$$

Once the Thévenin voltage and Norton current are calculated, the equivalent resistance between terminals a and b is obtained as follows:

$$R_{eq} = \frac{U_{th}}{I_{nor}} = \frac{8}{8/6} = 6\,\Omega$$

In this case, since the circuit has a dependent source, the equivalent resistance cannot be obtained by associating resistors in the passive circuit. Instead, a test source could be used connected between terminals a and b to the passive circuit, as shown in Fig. 2.72.

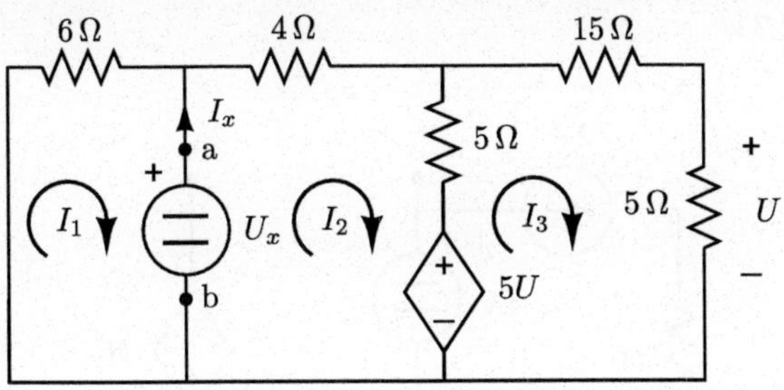

Fig. 2.72 Circuit with a test voltage source

According to Fig. 2.72, applying the mesh analysis method yields:

$$\begin{pmatrix} 6 & 0 & 0 \\ 0 & 4+5 & -5 \\ 0 & -5 & 5+15+5 \end{pmatrix} \begin{pmatrix} I_1 \\ I_2 \\ I_3 \end{pmatrix} = \begin{pmatrix} -U_x \\ U_x - 5U \\ 5U \end{pmatrix}$$

On the other hand:

$$U = 5 \cdot I_3$$

Therefore:

$$\begin{pmatrix} 6 & 0 & 0 \\ 0 & 9 & -5 \\ 0 & -5 & 25 \end{pmatrix} \begin{pmatrix} I_1 \\ I_2 \\ I_3 \end{pmatrix} = \begin{pmatrix} -U_x \\ U_x - 5 \cdot (5 \cdot I_3) \\ 5 \cdot (5 \cdot I_3) \end{pmatrix} = \begin{pmatrix} -U_x \\ U_x - 25 \cdot I_3 \\ 25 \cdot I_3 \end{pmatrix}$$

Rearranging terms:

$$\begin{pmatrix} 6 & 0 & 0 \\ 0 & 9 & 20 \\ 0 & -5 & 0 \end{pmatrix} \begin{pmatrix} I_1 \\ I_2 \\ I_3 \end{pmatrix} = \begin{pmatrix} -U_x \\ U_x \\ 0 \end{pmatrix}$$

and solving

$$I_1 = \frac{-1}{6} \cdot U_x \, \mathrm{A} \ ; \ I_2 = 0 \, \mathrm{A} \ ; \ I_3 = \frac{1}{20} \cdot U_x \, \mathrm{A}$$

Consequently, considering that $I_x = I_2 - I_1$:

$$R_{eq} = \frac{U_x}{I_x} = \frac{U_x}{0 - (-U_x/6)} = 6 \, \Omega.$$

2.23 Thévenin and Norton Equivalents

Determine the Thévenin and Norton equivalents between terminals a and b of the circuit in Fig. 2.73.

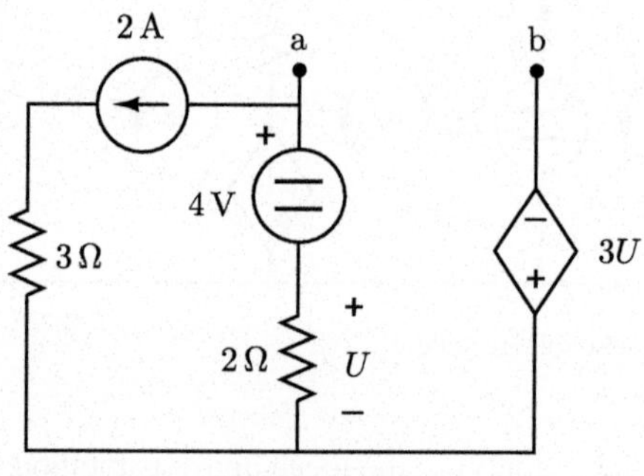

Fig. 2.73 Circuit considered in the problem under study

Solution First, we will calculate the open-circuit voltage between terminals a and b, the Thévenin voltage. To do this, we will solve the circuit shown in Fig. 2.74. According to it, applying Kirchhoff's voltage law yields:

$$U_{th} - 3U + 2 \cdot 2 - 4 = 0 \ \Rightarrow \ U_{th} = 3U$$

On the other hand:

$$U = -2 \cdot 2 = -4 \, \mathrm{V}$$

Therefore:

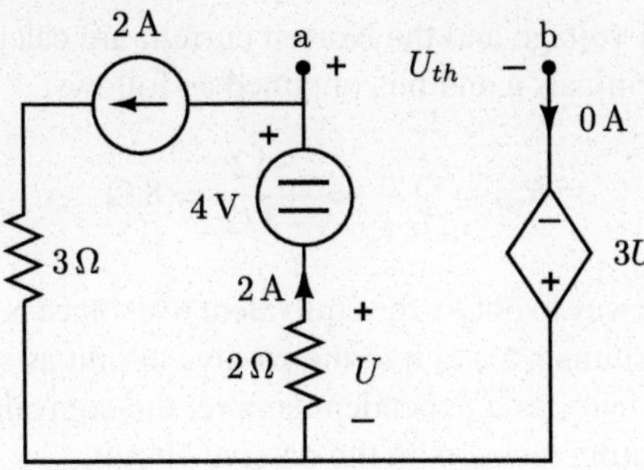

Fig. 2.74 Circuit considered in the problem under study

$$U_{th} = 3U = 3 \cdot (-4) = -12\,\text{V}$$

Next, we calculate the Norton current, which is obtained by short-circuiting terminals a and b, as shown in Fig. 2.75.

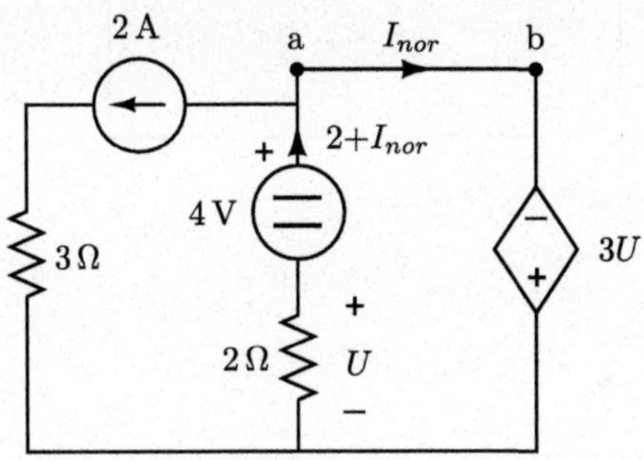

Fig. 2.75 Circuit with terminals a and b short-circuited

According to Fig. 2.75, applying Kirchhoff's voltage law yields:

$$-3U + 2 \cdot (2 + I_{nor}) - 4 = 0 \;\Rightarrow\; I_{nor} = \frac{3}{2} \cdot U$$

On the other hand:

$$U = -2 \cdot (2 + I_{nor})$$

Therefore:

$$I_{nor} = \frac{3}{2} \cdot U = \frac{3}{2} \cdot (-2 \cdot (2 + I_{nor})) \;\Rightarrow\; I_{nor} = \frac{-3}{2}\,\text{A}$$

Once the Thévenin voltage and the Norton current are calculated, the equivalent resistance between terminals a and b is obtained as follows:

$$R_{eq} = \frac{U_{th}}{I_{nor}} = \frac{-12}{-3/2} = 8\,\Omega$$

Another alternative way to obtain the equivalent resistance is by using a test source connected between terminals a and b to the passive circuit as shown in Fig. 2.76. In this case, since the circuit has a dependent source, the equivalent resistance cannot be obtained by associating resistors in the passive circuit.

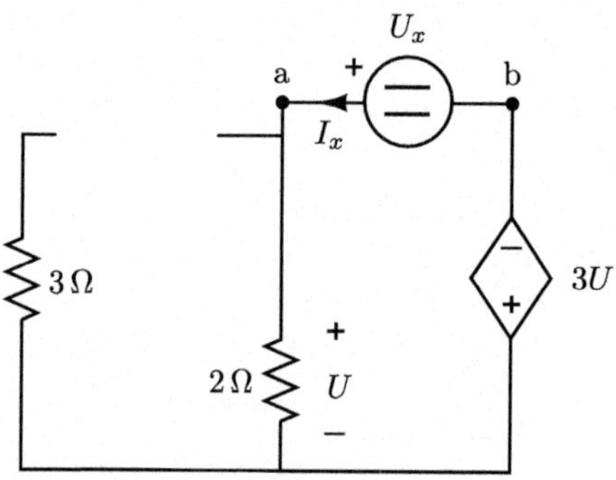

Fig. 2.76 Circuit with a test voltage source

According to Fig. 2.76:

$$U_x - 3U - 2 \cdot I_x = 0$$

On the other hand:

$$U = 2 \cdot I_x$$

Consequently:

$$U_x - 3 \cdot (2 \cdot I_x) - 2 \cdot I_x = 0 \;\Rightarrow\; U_x = 8 \cdot I_x \;\Rightarrow\; R_{eq} = \frac{U_x}{I_x} = 8\,\Omega.$$

2.24 Thévenin and Norton Equivalents

In the circuit of Fig. 2.77, determine the Thévenin and Norton equivalents between terminals a and b.

Solution First, we will calculate the open-circuit voltage between terminals a and b, the Thévenin voltage. To do this, we will solve the circuit shown in Fig. 2.78. According to it, applying Kirchhoff's voltage law yields:

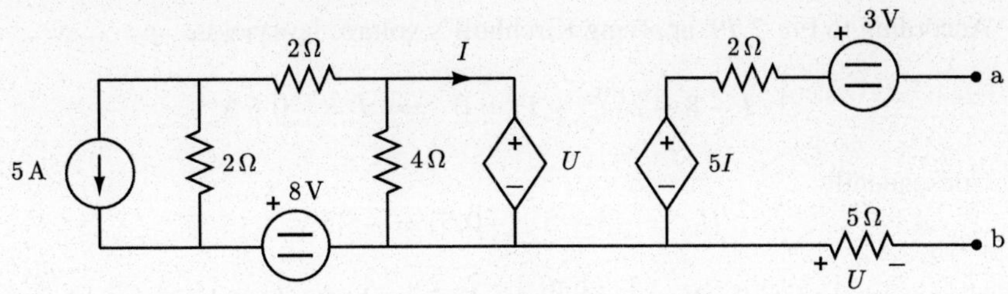

Fig. 2.77 Circuit considered in the problem under study

$$U_{th} = -3 + 5I$$

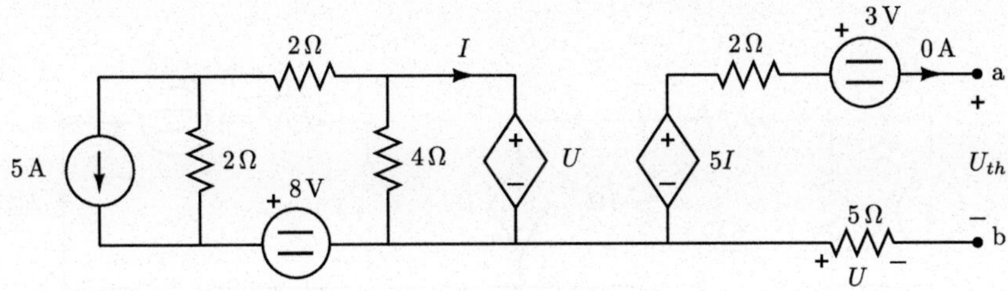

Fig. 2.78 Circuit considered in the problem under study

On the other hand, since terminals a and b are open-circuited, there is no current flowing through the 5 Ω resistor and, consequently:

$$U = 0\,\text{V}$$

To obtain the current I, we will analyze the left part of the circuit, which has been represented in Fig. 2.79. In this circuit, the dependent voltage source has been replaced by a short circuit since $U = 0\,\text{V}$.

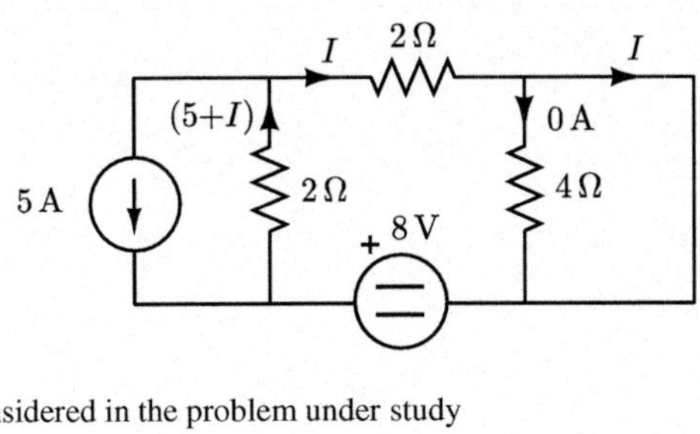

Fig. 2.79 Circuit considered in the problem under study

According to Fig. 2.79, applying Kirchhoff's voltage law yields:

$$2 \cdot I - 8 + 2 \cdot (5 + I) = 0 \;\Rightarrow\; I = -0{,}5\,\mathrm{A}$$

Consequently:

$$U_{th} = -3 + 5 \cdot (-0{,}5) = -5{,}5\,\mathrm{V}$$

Next, we calculate the Norton current, which is obtained by short-circuiting terminals a and b, as shown in Fig. 2.80. The left part of the circuit in Fig. 2.80 can be simplified, resulting in the circuit in Fig. 2.81. Applying the mesh analysis method:

$$\begin{pmatrix} 4 & 0 \\ 0 & 7 \end{pmatrix} \begin{pmatrix} I_1 \\ I_2 \end{pmatrix} = \begin{pmatrix} -U + 8 - 10 \\ -3 + 5I \end{pmatrix}$$

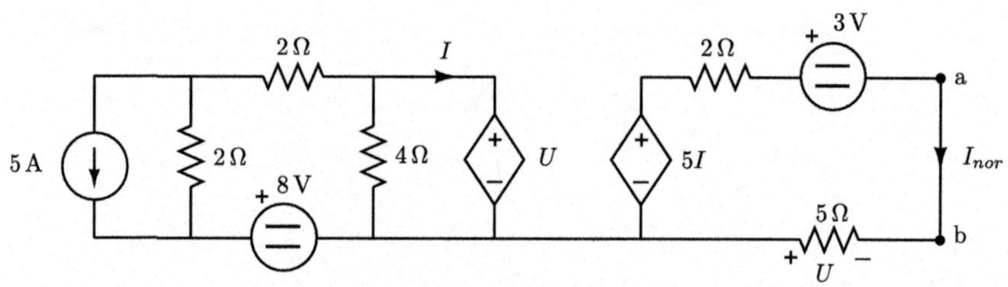

Fig. 2.80 Circuit with terminals a and b short-circuited

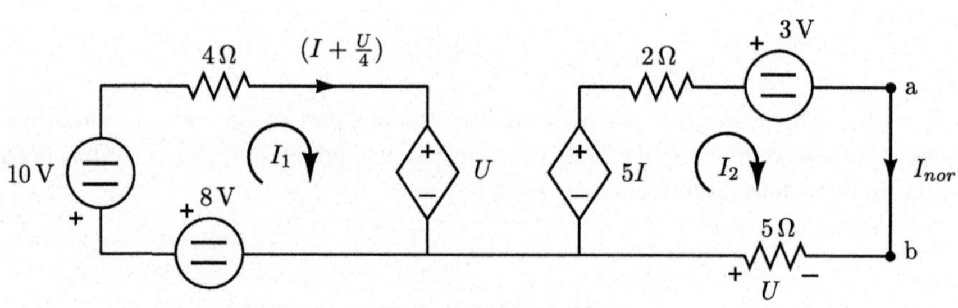

Fig. 2.81 Simplified circuit

On the other hand:

$$U = -5 \cdot I_2 \;;\; I = I_1 - \frac{U}{4} = I_1 + \frac{5 \cdot I_2}{4}$$

Substituting values:

$$\begin{pmatrix} 4 & 0 \\ 0 & 7 \end{pmatrix} \begin{pmatrix} I_1 \\ I_2 \end{pmatrix} = \begin{pmatrix} 5 \cdot I_2 - 2 \\ -3 + 5 \cdot (I_1 + \frac{5 \cdot I_2}{4}) \end{pmatrix} = \begin{pmatrix} 5 \cdot I_2 - 2 \\ -3 + 5 \cdot I_1 + \frac{25 \cdot I_2}{4} \end{pmatrix}$$

rearranging terms:

$$\begin{pmatrix} 4 & -5 \\ -5 & \frac{3}{4} \end{pmatrix} \begin{pmatrix} I_1 \\ I_2 \end{pmatrix} = \begin{pmatrix} -2 \\ -3 \end{pmatrix}$$

and solving, we obtain the current I_2, which is the same as the Norton current:

$$I_{nor} = I_2 = 1 \text{ A}$$

Once the Thévenin voltage and the Norton current are obtained, the equivalent resistance between terminals a and b is calculated as follows:

$$R_{eq} = \frac{U_{th}}{I_{nor}} = \frac{-5{,}5}{1} = -5{,}5\ \Omega$$

It can be observed that the equivalent resistance has a negative value, which is due to the presence of dependent sources in the circuit.

Another way to calculate the equivalent resistance is from the passive circuit. In this case, since the circuit has a dependent source, the equivalent resistance cannot be obtained by associating resistors. Instead, a test source could be used connected between terminals a and b to the passive circuit, as shown in Fig. 2.82.

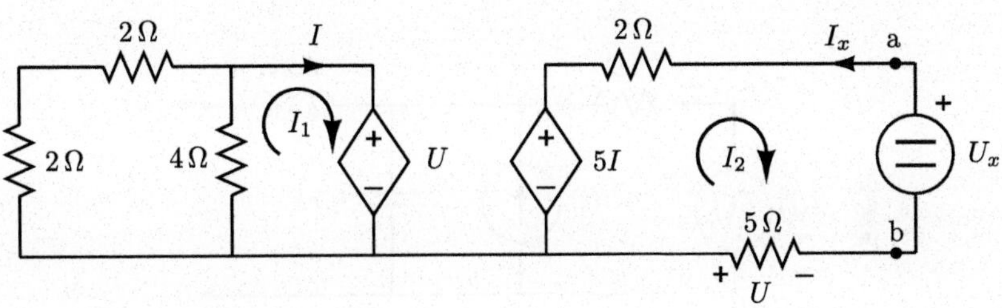

Fig. 2.82 Circuit with a test voltage source

According to Fig. 2.82, applying the mesh analysis method:

$$\begin{pmatrix} 2 & 0 \\ 0 & 7 \end{pmatrix} \begin{pmatrix} I_1 \\ I_2 \end{pmatrix} = \begin{pmatrix} -U \\ 5I - U_x \end{pmatrix}$$

On the other hand:

$$U = -5 \cdot I_2 \ ; \ \ I = I_1$$

Substituting values:

$$\begin{pmatrix} 2 & 0 \\ 0 & 7 \end{pmatrix} \begin{pmatrix} I_1 \\ I_2 \end{pmatrix} = \begin{pmatrix} 5 \cdot I_2 \\ 5 \cdot I_1 - U_x \end{pmatrix}$$

rearranging terms:

$$\begin{pmatrix} 2 & -5 \\ -5 & 7 \end{pmatrix} \begin{pmatrix} I_1 \\ I_2 \end{pmatrix} = \begin{pmatrix} 0 \\ -U_x \end{pmatrix}$$

and solving, we obtain the mesh currents I_1 and I_2:

$$I_1 = \frac{5}{11} \cdot U_x \text{ A} \; ; \; I_2 = \frac{2}{11} \cdot U_x \text{ A}$$

As $I_x = -I_2$ then the relationship between U_x and I_x is as follows:

$$R_{eq} = \frac{U_x}{I_x} = \frac{U_x}{-I_2} = \frac{U_x}{-\dfrac{2}{11} \cdot U_x} = -5{,}5\,\Omega.$$

2.25 Thévenin and Norton Equivalents

In the circuit of Fig. 2.83, calculate the Thévenin and Norton equivalents between terminals a and b, knowing that $i(t) = 9\cos(10t - 60°)$ A and $e(t) = 9\cos(10t)$ V.

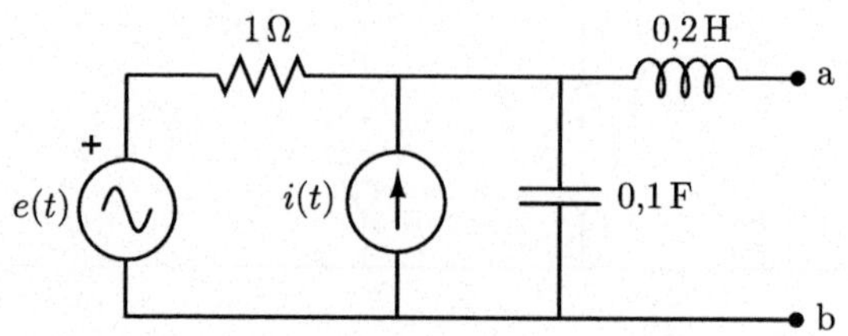

Fig. 2.83 Circuit considered in the problem under study

Solution First, the circuit will be transformed into the phasor domain. To do this, the sources are replaced by their corresponding phasors, and the resistor, inductor, and capacitor are replaced by their impedance, resulting in the circuit shown in Fig. 2.84.

The Thévenin voltage is obtained from the circuit in Fig. 2.84 by leaving terminals a and b open. Applying the nodal analysis method, we get:

$$\left(\frac{1}{1} + \frac{1}{-j}\right) \cdot \overline{U}_{th} = \frac{9/\sqrt{2}\angle 0°}{1} + \frac{9}{\sqrt{2}}\angle - 60° \;\Rightarrow\; \overline{U}_{th} \approx 7{,}79\angle - 75°\text{ V}$$

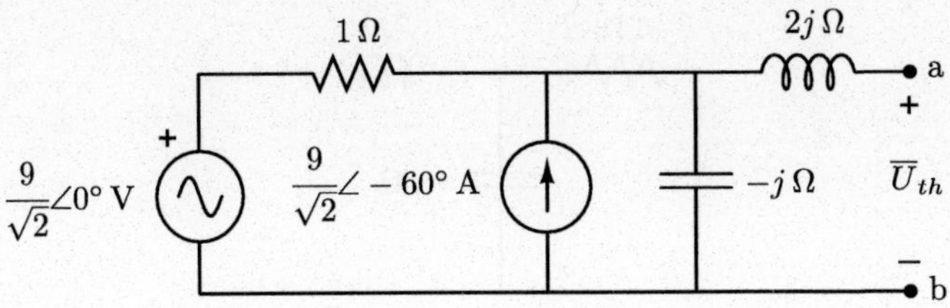

Fig. 2.84 Modification in the original circuit

Next, the Norton current is calculated, which is obtained by short-circuiting terminals a and b, as shown in Fig. 2.85. It can be observed that the voltage $\overline{U}$ has been introduced into the circuit in order to apply nodal analysis, resulting in:

$$\left(\frac{1}{1}+\frac{1}{-j}+\frac{1}{2j}\right)\cdot\overline{U}=\frac{9/\sqrt{2}\angle 0^\circ}{1}+\frac{9}{\sqrt{2}}\angle-60^\circ \;\Rightarrow\; \overline{U}\approx 9{,}86\angle-56{,}57^\circ\text{ V}$$

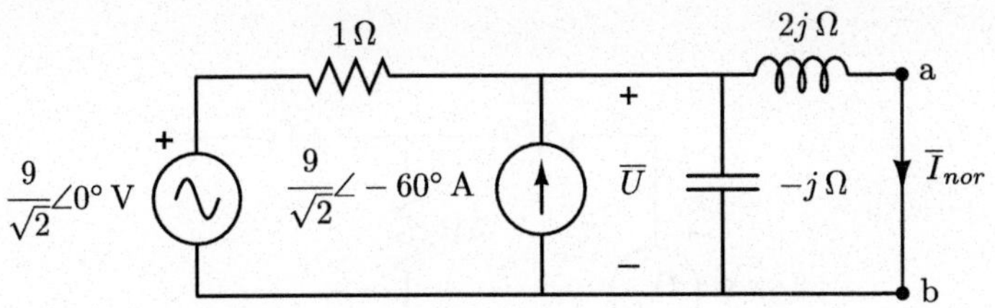

Fig. 2.85 Circuit with terminals a and b short-circuited

Once the voltage $\overline{U}$ is obtained, the Norton current is as follows:

$$\overline{I}_{nor}=\frac{\overline{U}}{2j}=\frac{9{,}86\angle-56{,}57^\circ}{2j}=4{,}93\angle-146{,}57^\circ\text{ A}$$

After calculating the Thévenin voltage and the Norton current, the equivalent impedance between terminals a and b is obtained as follows:

$$\overline{Z}_{eq}=\frac{\overline{U}_{th}}{\overline{I}_{nor}}=\frac{7{,}79\angle-75^\circ}{4{,}93\angle-146{,}57^\circ}\approx 1{,}58\angle 71{,}57^\circ\ \Omega$$

The equivalent impedance could also have been obtained by nullifying the sources in the circuit of Fig. 2.84, as shown in Fig. 2.86.

According to Fig. 2.86, the equivalent impedance is calculated as follows:

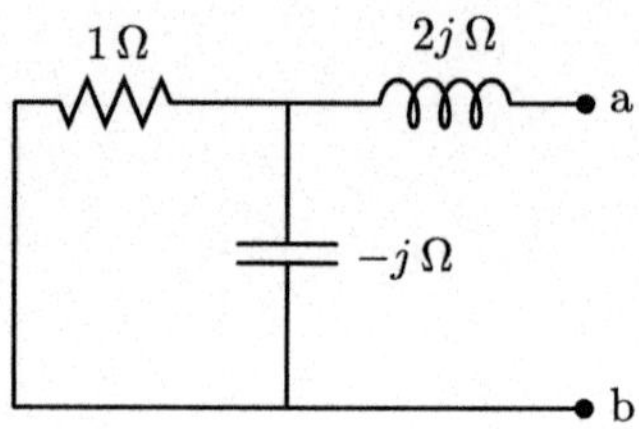

Fig. 2.86 Passive circuit

$$\overline{Z}_{eq} = 2j + \frac{1 \cdot (-j)}{1 - j} \approx 1{,}58\angle 71{,}57^\circ\ \Omega.$$

2.26 Thévenin and Norton Equivalents

Find the Thévenin and Norton equivalents between terminals a and b of the circuit in Fig. 2.87.

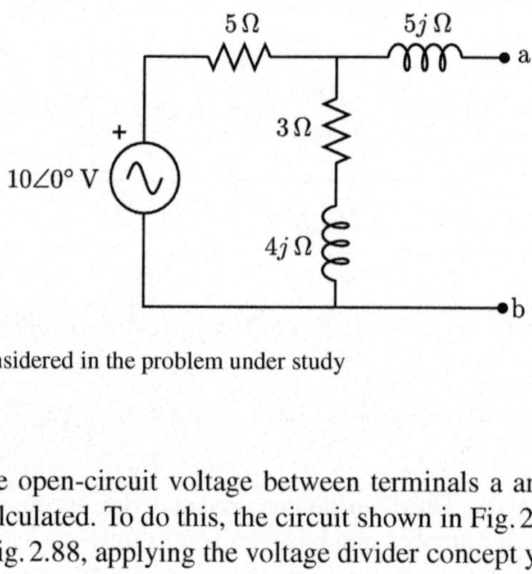

Fig. 2.87 Circuit considered in the problem under study

Solution First, the open-circuit voltage between terminals a and b, the Thévenin voltage, will be calculated. To do this, the circuit shown in Fig. 2.88 will be solved.

According to Fig. 2.88, applying the voltage divider concept yields:

$$\overline{U}_{th} = 10\angle 0^\circ \cdot \frac{3 + 4j}{3 + 4j + 5} \approx 5{,}59\angle 26{,}57^\circ\ \mathrm{V}$$

Next, the Norton current is calculated, which is obtained by short-circuiting terminals a and b, as shown in Fig. 2.89. It can be observed that the mesh currents $\overline{I}_1$ and $\overline{I}_2$ have been introduced into the circuit in order to apply the mesh analysis method, resulting in:

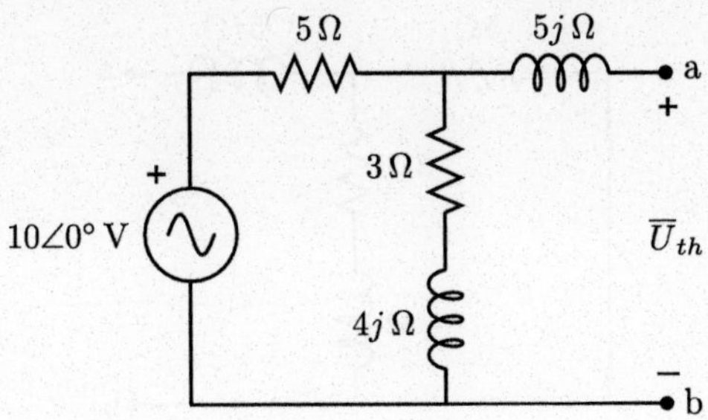

Fig. 2.88 Circuit considered in the problem under study

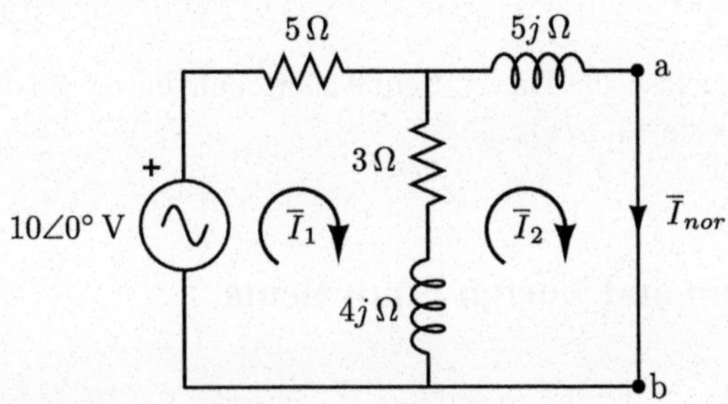

Fig. 2.89 Circuit with terminals a and b short-circuited

$$\begin{pmatrix} 5+3+4j & -3-4j \\ -3-4j & 3+4j+5j \end{pmatrix} \begin{pmatrix} \overline{I}_1 \\ \overline{I}_2 \end{pmatrix} = \begin{pmatrix} 10\angle 0^\circ \\ 0 \end{pmatrix}$$

Solving:

$$\overline{I}_1 = 1{,}58\angle -23{,}20^\circ \text{ A} \ ; \ \overline{I}_2 = 0{,}83\angle -41{,}63^\circ \text{ A}$$

Consequently:

$$\overline{I}_{nor} = \overline{I}_2 = 0{,}83\angle -41{,}63^\circ \text{ A}$$

After calculating the Thévenin voltage and the Norton current, the equivalent impedance between terminals a and b is obtained as follows:

$$\overline{Z}_{eq} = \frac{\overline{U}_{th}}{\overline{I}_{nor}} = \frac{5{,}59\angle 26{,}57^\circ}{0{,}83\angle -41{,}63^\circ} = 6{,}73\angle 68{,}2^\circ \ \Omega$$

The equivalent impedance could also have been obtained by making the circuit of Fig. 2.87 passive, as shown in Fig. 2.90.

According to Fig. 2.90, the equivalent impedance is calculated as follows:

$$\overline{Z}_{eq} = 5j + \frac{5 \cdot (3+4j)}{5+3+4j} \approx 6{,}73\angle 68{,}2^\circ \ \Omega$$

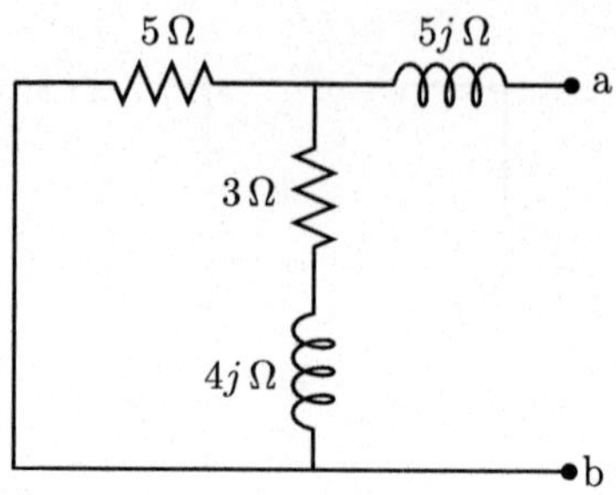

Fig. 2.90 Passive circuit

In this particular case, the Thévenin equivalent could have been obtained by source transformation, as shown in Fig. 2.91.

2.27 Thévenin and Norton Equivalents

In the circuit of Fig. 2.92, the impedance $\overline{Z}_L$ absorbs 9/8 W and 9/8 var. It is also known that $\overline{Z}_L$ is the impedance that absorbs the maximum power of all those that can be connected between a and b. Calculate the Thévenin and Norton equivalents between a and b (with $\overline{Z}_L$ disconnected) and the value of $\overline{Z}_4$.

Solution First, we will calculate the open-circuit voltage between terminals a and b, Thévenin voltage. To do this, we will solve the circuit shown in Fig. 2.93. In this circuit, each real voltage source has been replaced by its equivalent current source, and the series connection of the current source with impedance $\overline{Z}_1$ has been replaced by the equivalent current source.

According to Fig. 2.99, applying the node method yields:

$$\left(\frac{1}{2\angle 0^\circ}+\frac{1}{2\angle 0^\circ}+\frac{1}{\overline{Z}_4}+\frac{1}{\overline{Z}_4}+\frac{1}{\overline{Z}_4}\right)\cdot\overline{U}_{th}=2\angle 0^\circ+1\angle 0^\circ+\frac{\overline{E}_1}{\overline{Z}_4}-\frac{\overline{E}_1}{\overline{Z}_4}$$

Solving for Thévenin voltage:

$$\overline{U}_{th}=\frac{3\cdot\overline{Z}_4}{3+\overline{Z}_4}$$

Next, we calculate the Norton current, which is obtained by short-circuiting terminals a and b, as shown in Fig. 2.94.

According to Fig. 2.94, it is easy to verify that all impedances are short-circuited and, therefore, no current flows through them. Consequently:

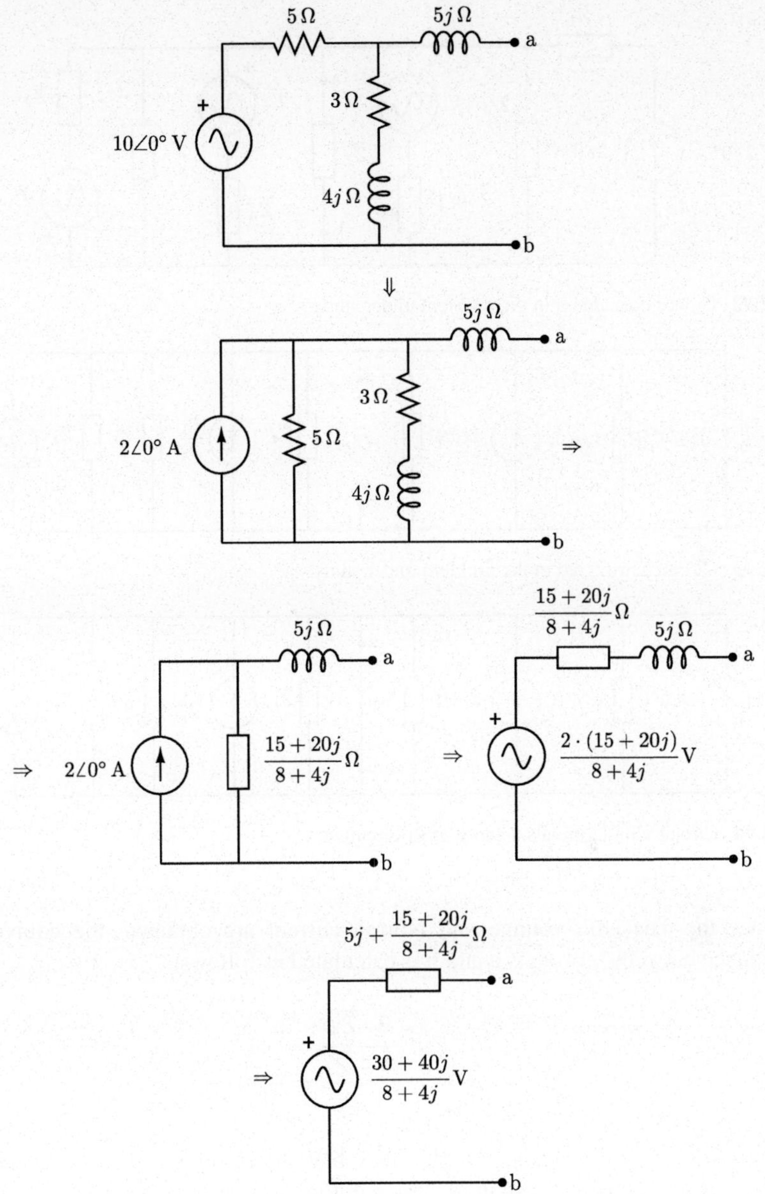

Fig. 2.91 Modifications in the original circuit

$$\overline{I}_{nor} = 2\angle 0^\circ + 1\angle 0^\circ + \frac{\overline{E}_1}{\overline{Z}_4} - \frac{\overline{E}_1}{\overline{Z}_4} = 3\angle 0^\circ \text{ A}$$

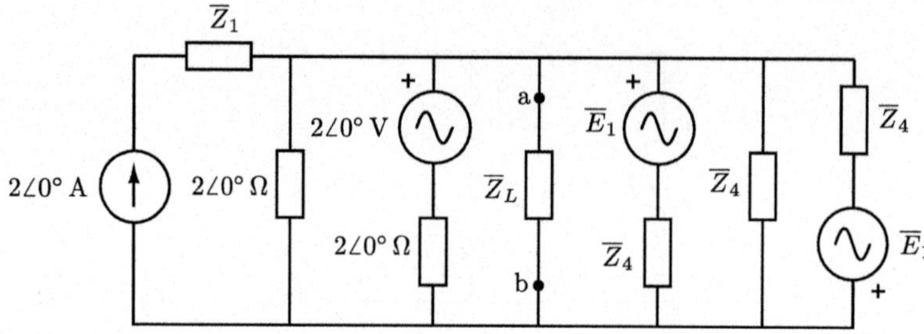

Fig. 2.92 Circuit considered in the problem under study

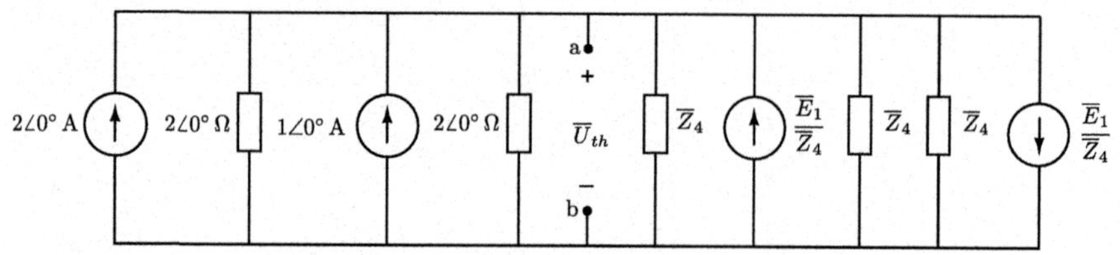

Fig. 2.93 Circuit considered in the problem under study

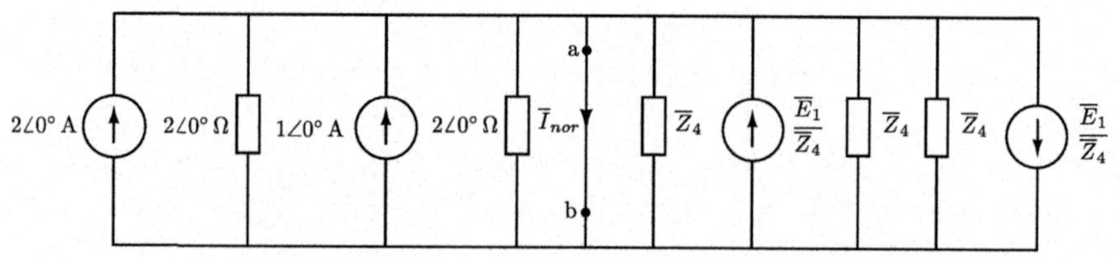

Fig. 2.94 Circuit with terminals a and b short-circuited

Once the Thévenin voltage and Norton current are obtained, the equivalent impedance between terminals a and b is calculated as follows:

$$\overline{Z}_{eq} = \frac{\overline{U}_{th}}{\overline{I}_{nor}} = \frac{\frac{3 \cdot \overline{Z}_4}{3 + \overline{Z}_4}}{3\angle 0°} = \frac{\overline{Z}_4}{3 + \overline{Z}_4}$$

The equivalent impedance could also have been obtained by associating impedances in the passive circuit as shown in Fig. 2.95.

According to Fig. 2.95, it can be observed that all impedances are connected in parallel and therefore:

$$\overline{Z}_{eq} = \frac{1}{\frac{1}{2\angle 0°} + \frac{1}{2\angle 0°} + \frac{1}{\overline{Z}_4} + \frac{1}{\overline{Z}_4} + \frac{1}{\overline{Z}_4}} = \frac{\overline{Z}_4}{3 + \overline{Z}_4}$$

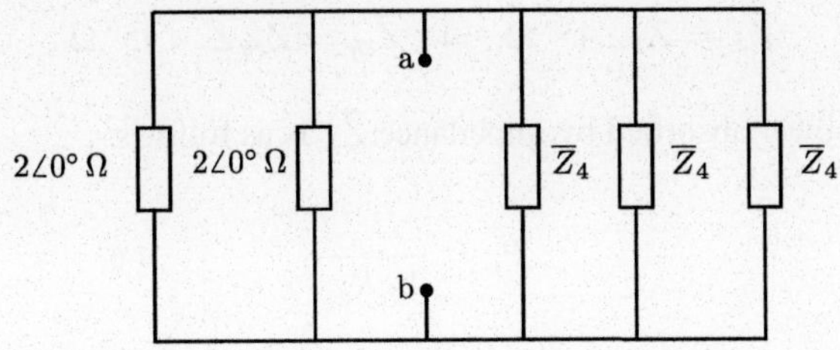

Fig. 2.95 Passive circuit

Figure 2.96 shows the Thévenin equivalent circuit between terminals a and b and the connected impedance $\overline{Z}_L$, where

$$\overline{U}_{th} = \frac{3\overline{Z}_4}{3+\overline{Z}_4} \ ; \ \overline{Z}_{eq} = \frac{\overline{Z}_4}{3+\overline{Z}_4}$$

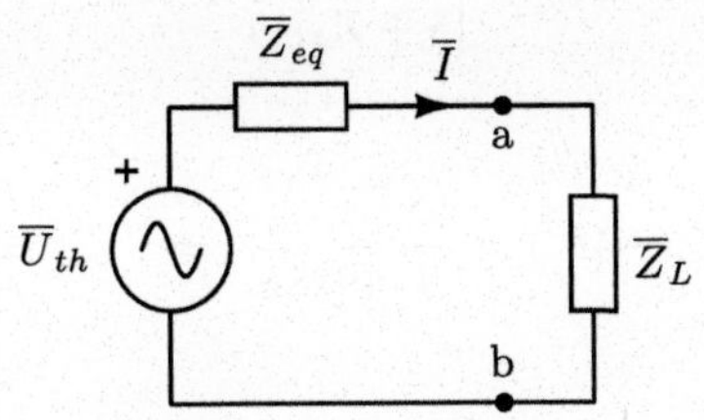

Fig. 2.96 Circuit considered in the problem under study

From Fig. 2.96:

$$\overline{I} = \frac{\overline{U}_{th}}{\overline{Z}_{eq} + \overline{Z}_L}$$

According to the statement, impedance $\overline{Z}_L$ absorbs the maximum power from terminals a and b of the circuit, therefore:

$$\overline{Z}_L = \overline{Z}_{eq}^* = R_{eq} - jX_{eq}$$

Thus:

$$\overline{I} = \frac{\overline{U}_{th}}{\overline{Z}_{eq} + \overline{Z}_L} = \frac{\overline{U}_{th}}{\overline{Z}_{eq} + \overline{Z}_{eq}^*} = \frac{\overline{U}_{th}}{2 \cdot R_{eq}}$$

On the other hand, according to the statement, the active and reactive power absorbed by impedance $\overline{Z}_L$ are the same. This means that $R_L = X_L$ and consequently:

$$\overline{Z}_L = Z_L \angle 45^\circ\ \Omega \ \Rightarrow\ \overline{Z}_{eq} = Z_{eq} \angle -45^\circ\ \Omega$$

The active power absorbed by impedance $\overline{Z}_L$ is as follows:

$$P_L = \frac{U_{th}^2}{4 \cdot R_{eq}}$$

This power is, according to the statement, 9/8 W, therefore:

$$P_L = \frac{U_{th}^2}{4 \cdot R_{eq}} = \frac{9}{8} \ \Rightarrow\ R_{eq} = \frac{U_{th}^2}{4 \cdot (9/8)} = \frac{U_{th}^2}{4 \cdot (9/8)}$$

Taking into account that

$$R_{eq} = Z_{eq} \cdot \cos(-45^\circ)$$

and that

$$U_{th} = \frac{3 \cdot \left|\overline{Z}_4\right|}{\left|3 + \overline{Z}_4\right|} = 3 \cdot Z_{eq}$$

then

$$Z_{eq} \cdot \cos(-45^\circ) = \frac{\left(3 \cdot Z_{eq}\right)^2}{4 \cdot (9/8)} \ \Rightarrow\ Z_{eq} = \frac{\sqrt{2}}{4}\ \Omega$$

Therefore:

$$\overline{Z}_{eq} = Z_{eq} \angle -45^\circ = \frac{\sqrt{2}}{4} \angle -45^\circ = 0{,}25 - 0{,}25j\ \Omega$$

Once the equivalent impedance is calculated, impedance $\overline{Z}_4$ is calculated as follows:

$$\overline{Z}_{eq} = \frac{\overline{Z}_4}{3 + \overline{Z}_4} \ \Rightarrow\ \overline{Z}_4 = \frac{3 \cdot \overline{Z}_{eq}}{1 - \overline{Z}_{eq}}$$

Substituting values yields:

$$\overline{Z}_4 = \frac{3 \cdot \overline{Z}_{eq}}{1 - \overline{Z}_{eq}} = \frac{3 \cdot (0{,}25 - 0{,}25j)}{1 - (0{,}25 - 0{,}25j)} \approx 1{,}341 \angle -63{,}43^\circ\ \Omega$$

Finally, the Thévenin voltage is as follows:

$$\overline{U}_{th} = \frac{3 \cdot \overline{Z}_4}{3 + \overline{Z}_4} = 3 \cdot \overline{Z}_{eq} = 3 \cdot (0{,}25 - 0{,}25j) \approx 1{,}06 \angle -45^\circ\ \text{V}$$

Another alternative way to obtain impedance $\overline{Z}_4$ is by using the Norton equivalent from terminals a and b of Fig. 2.97.

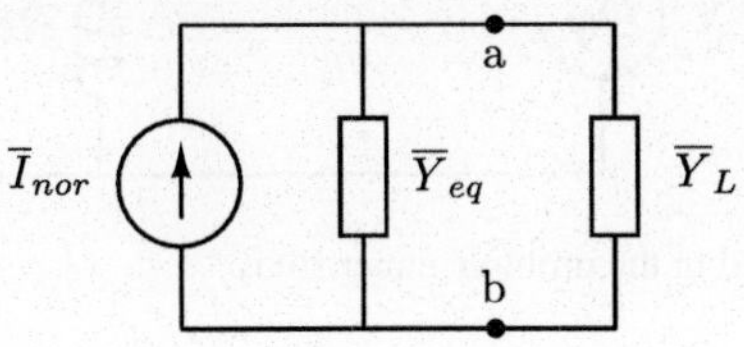

Fig. 2.97 Circuit considered in the problem under study

According to the statement, impedance $\overline{Z}_L$ absorbs the maximum power from terminals a and b of the circuit, therefore:

$$\overline{Y}_L = \frac{1}{\overline{Z}_L} = \frac{1}{\overline{Z}^*_{eq}} = \overline{Y}^*_{eq} = G_{eq} - jB_{eq}$$

The active power absorbed by admittance $\overline{Y}_L$ is as follows:

$$P_L = \frac{I^2_{nor}}{4 \cdot G_{eq}}$$

This power is, according to the statement, 9/8 W, therefore:

$$P_L = \frac{I^2_{nor}}{4 \cdot G_{eq}} = \frac{9}{8} \Rightarrow G_{eq} = \frac{3^2}{4 \cdot (9/8)} = 2\,\text{S}$$

On the other hand, according to the statement, the active and reactive power absorbed by impedance $\overline{Z}_L$ are the same. In terms of admittance, this means that $G_{eq} = B_{eq}$. Consequently:

$$\overline{Y}_{eq} = G_{eq} + jB_{eq} = 2 + 2j\,\text{S}$$

Once the equivalent admittance is calculated, impedance $\overline{Z}_4$ is calculated as follows:

$$\overline{Y}_{eq} = \frac{3 + \overline{Z}_4}{\overline{Z}_4} \Rightarrow \overline{Z}_4 = \frac{3}{1 + 2j} \approx 1{,}341\angle - 63{,}43°\,\Omega.$$

2.28 Thévenin and Norton Equivalents

Determine the Thévenin and Norton equivalents, seen from terminals a and b, of the circuit in Fig. 2.98.

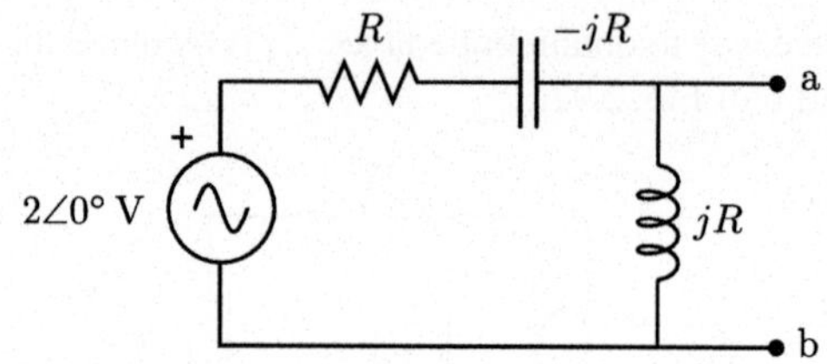

Fig. 2.98 Circuit considered in the problem under study

Solution First, the open-circuit voltage between terminals a and b, Thévenin voltage, will be calculated. To do this, the circuit shown in Fig. 2.99 will be solved. Applying the voltage divider concept results in:

$$\overline{U}_{th} = 2\angle 0^\circ \cdot \frac{jR}{R - jR + jR} = 2\angle 90^\circ \text{ V}$$

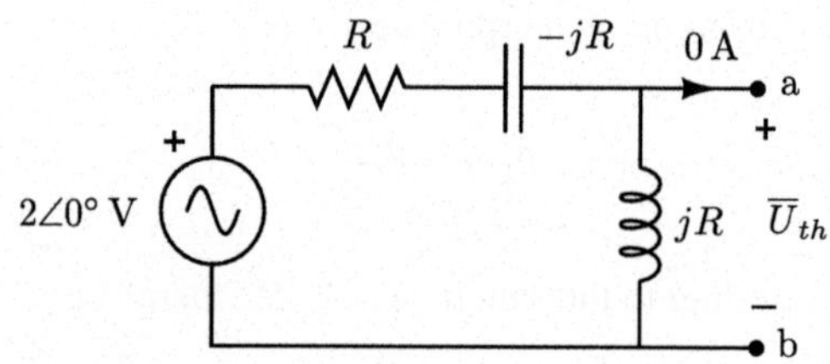

Fig. 2.99 Circuit considered in the problem under study

Next, the Norton current is calculated, which is obtained by short-circuiting terminals a and b, as shown in Fig. 2.100.

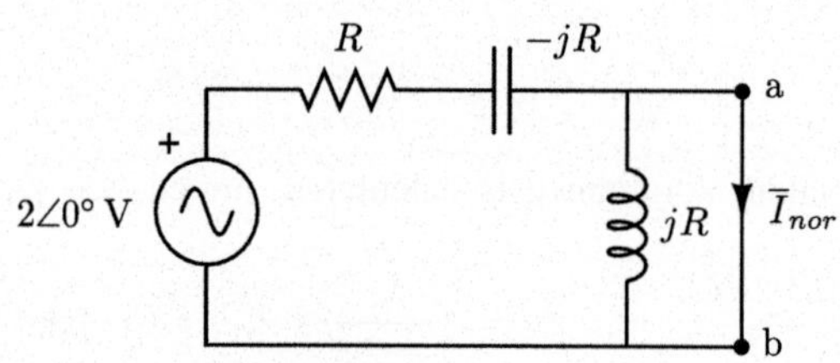

Fig. 2.100 Circuit with terminals a and b short-circuited

According to Fig. 2.100:

$$\overline{I}_{nor} = \frac{2\angle 0^\circ}{R - jR} = \frac{\sqrt{2}}{R}\angle 45^\circ \text{ A}$$

Once the Thévenin voltage and the Norton current are obtained, the equivalent impedance between terminals a and b is calculated as follows:

$$\overline{Z}_{eq} = \frac{\overline{U}_{th}}{\overline{I}_{nor}} = \frac{2\angle 90^\circ}{\frac{\sqrt{2}}{R}\angle 45^\circ} = R\sqrt{2}\angle 45^\circ\ \Omega$$

The equivalent impedance could also have been obtained by associating impedances in the passive circuit as shown in Fig. 2.101.

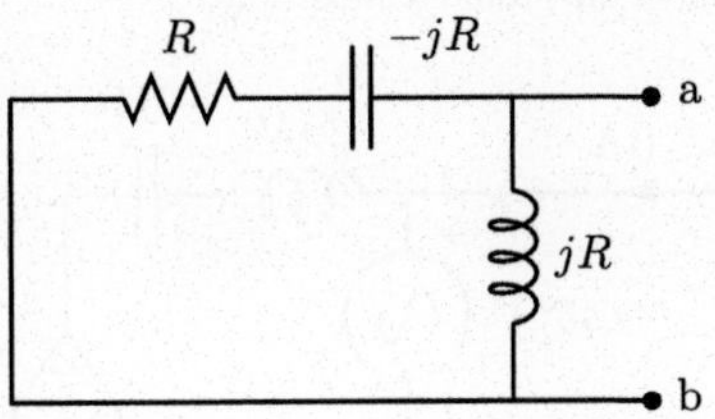

Fig. 2.101 Passive circuit

According to Fig. 2.101:

$$\overline{Z}_{eq} = (R - jR)//jR = \frac{(R - jR) \cdot jR}{R - jR + jR} = \frac{R\sqrt{2}\angle -45^\circ \cdot R\angle 90^\circ}{R} = R\sqrt{2}\angle 45^\circ\ \Omega.$$

2.29 Thévenin and Norton Equivalents

In the circuit of Fig. 2.102, determine the Thévenin and Norton equivalents from terminals a and b.

Solution First, the open-circuit voltage between terminals a and b, Thévenin voltage, will be calculated. To do this, the circuit shown in Fig. 2.103 will be solved.

According to Fig. 2.103:

$$\overline{U}_{th} = 4\angle 45^\circ - 2 \cdot \overline{I}$$

On the other hand:

$$(6j - 2j) \cdot \overline{I} - 6\overline{U} + 2 \cdot \overline{I} - 4\angle 45^\circ = 0$$

Additionally:

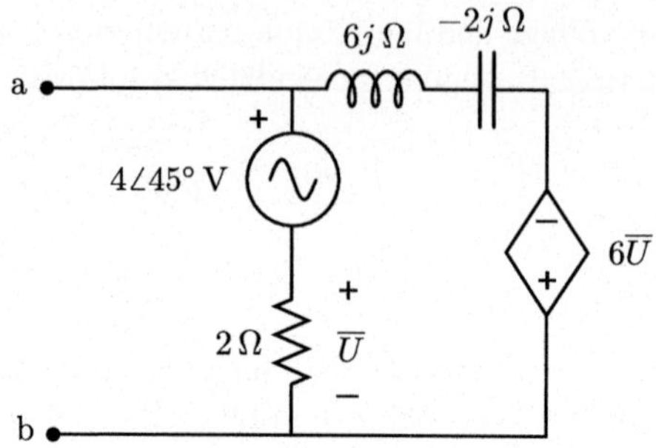

Fig. 2.102 Circuit considered in the problem under study

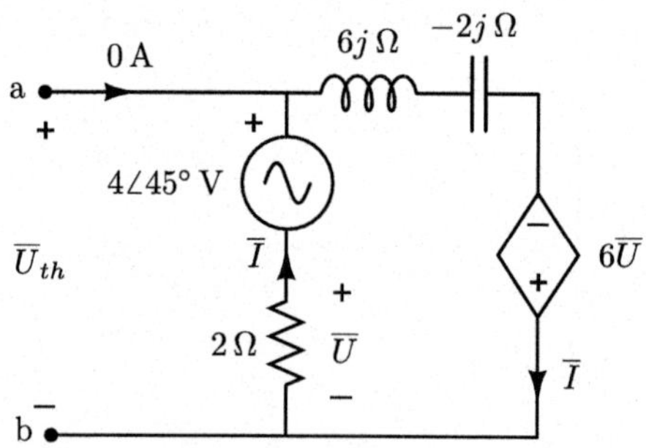

Fig. 2.103 Circuit considered in the problem under study

$$\overline{U} = -2 \cdot \overline{I}$$

Consequently:

$$(6j - 2j) \cdot \overline{I} + 12\overline{I} + 2 \cdot \overline{I} - 4\angle 45^\circ = 0 \Rightarrow \overline{I} = \frac{4\angle 45^\circ}{14 + 4j} \text{ A}$$

Thus, the Thévenin voltage is:

$$\overline{U}_{th} = 4\angle 45^\circ - 2 \cdot \frac{4\angle 45^\circ}{14 + 4j} \approx 3{,}47\angle 47{,}49^\circ \text{ V}$$

Next, the Norton current is calculated by short-circuiting terminals a and b, as shown in Fig. 2.104.

According to Fig. 2.104, the Norton current is obtained by applying Kirchhoff's current law:

$$\overline{I}_{nor} = -\overline{I}_1 - \overline{I}_2$$

Next, the currents $\overline{I}_1$ and $\overline{I}_2$ are calculated. According to Fig. 2.104:

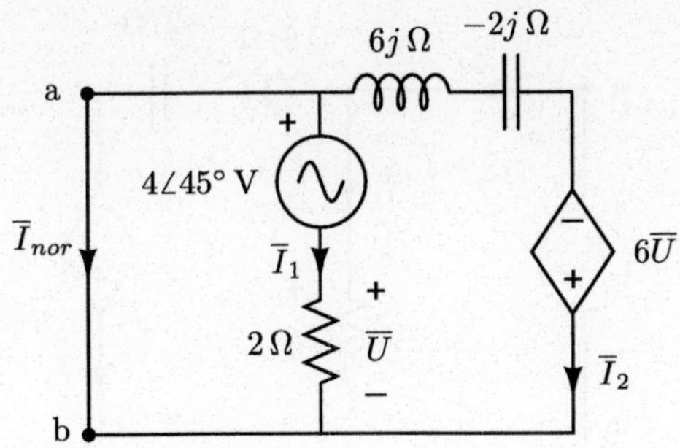

Fig. 2.104 Circuit with terminals a and b short-circuited

$$4\angle 45° + 2\cdot \overline{I}_1 = 0 \;\Rightarrow\; \overline{I}_1 = \frac{-4\angle 45°}{2} = 2\angle -135° \text{ A}$$

On the other hand:

$$(6j - 2j)\cdot \overline{I}_2 - 6\overline{U} = 0 \;\Rightarrow\; \overline{I}_2 = \frac{6\overline{U}}{4j}$$

Additionally:

$$\overline{U} = 2\cdot \overline{I}_1$$

Consequently:

$$\overline{I}_2 = \frac{6\cdot 2\cdot \overline{I}_1}{4j} = \frac{6\cdot 2\cdot (2\angle -135°)}{4j} = 6\angle 135° \text{ A}$$

Thus, the Norton current is:

$$\overline{I}_{nor} = -2\angle -135° - 6\angle 135° \approx 6{,}32\angle -26{,}57° \text{ A}$$

With the Thévenin voltage and the Norton current calculated, the equivalent impedance between terminals a and b is obtained as follows:

$$\overline{Z}_{eq} = \frac{\overline{U}_{th}}{\overline{I}_{nor}} = \frac{3{,}47\angle 47{,}49°}{6{,}32\angle -26{,}57°} \approx 0{,}55\angle 74°\;\Omega$$

Another way to obtain the equivalent impedance is by using a test source connected between terminals a and b to the passive circuit as shown in Fig. 2.105. In this case, as the circuit has a dependent source, the equivalent impedance cannot be obtained by impedance association in the passive circuit.

According to Fig. 2.105:

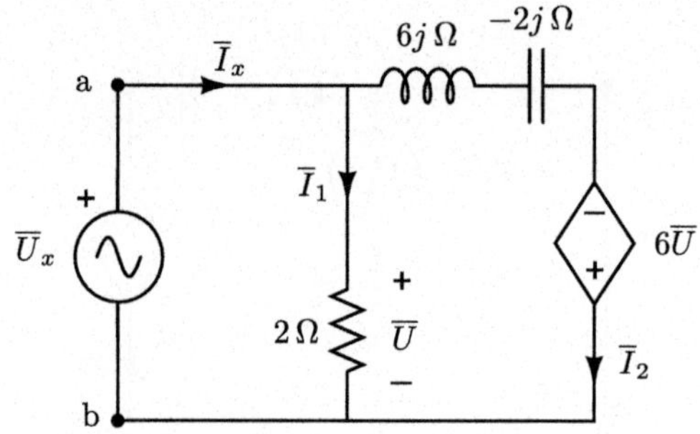

Fig. 2.105 Circuit with a test voltage source

$$\overline{I}_1 = \frac{\overline{U}_x}{2}$$

On the other hand:

$$\overline{U}_x = (6j - 2j) \cdot \overline{I}_2 - 6\overline{U}$$

Additionally:

$$\overline{U} = \overline{U}_x$$

Consequently:

$$\overline{U}_x = 4j \cdot \overline{I}_2 - 6 \cdot \overline{U}_x \;\Rightarrow\; \overline{I}_2 = \frac{7 \cdot \overline{U}_x}{4j}$$

Applying Kirchhoff's current law:

$$\overline{I}_x = \overline{I}_1 + \overline{I}_2 = \frac{\overline{U}_x}{2} + \frac{7 \cdot \overline{U}_x}{4j} = \frac{14+4j}{8j} \cdot \overline{U}_x$$

$$\Rightarrow\; \overline{Z}_{eq} = \frac{\overline{U}_x}{\overline{I}_x} = \frac{8j}{14+4j} \approx 0,55\angle 74^\circ\,\Omega.$$

2.30 Thévenin and Norton Equivalents

In the circuit of Fig. 2.106, determine the Thévenin and Norton equivalents between terminals a and b.

Solution First, the coupled coils will be replaced by their representation with dependent voltage sources in the phasor domain as shown in Fig. 2.107.

Next, the open-circuit voltage between terminals a and b, Thévenin voltage, will be calculated. To do this, the circuit shown in Fig. 2.108 will be solved. According

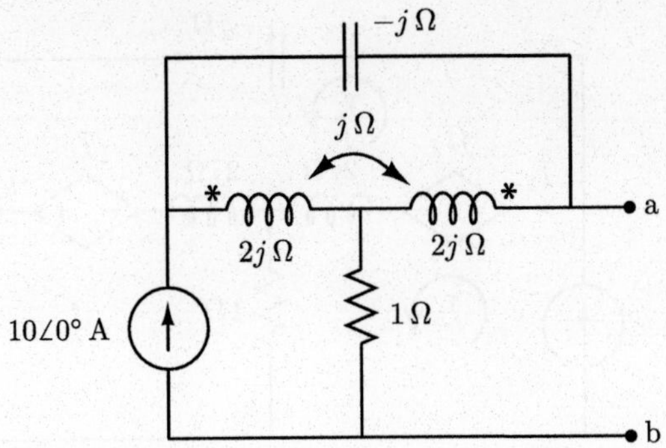

Fig. 2.106 Circuit considered in the problem under study

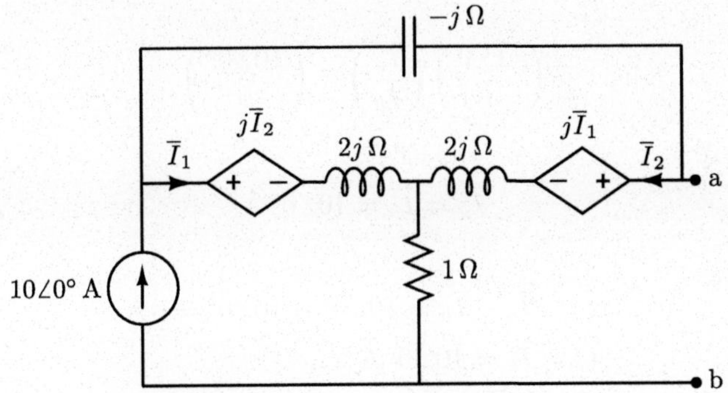

Fig. 2.107 Circuit with the equivalent model of coupled coils

to Fig. 2.108:

$$\overline{U}_{th} = j \cdot \overline{I}_1 + 2j \cdot \overline{I}_2 + 1 \cdot (\overline{I}_1 + \overline{I}_2)$$

By applying the mesh analysis, the currents $\overline{I}_a$ and $\overline{I}_b$ are calculated:

$$\begin{pmatrix} 1 & 0 \\ -2j & 2j + 2j - j \end{pmatrix} \begin{pmatrix} \overline{I}_a \\ \overline{I}_b \end{pmatrix} = \begin{pmatrix} 10\angle 0^\circ \\ j\overline{I}_2 - j\overline{I}_1 \end{pmatrix}$$

On the other hand:

$$\overline{I}_1 = \overline{I}_a - \overline{I}_b \ ; \ \overline{I}_2 = \overline{I}_b$$

Substituting values

$$\begin{pmatrix} 1 & 0 \\ -2j & 2j + 2j - j \end{pmatrix} \begin{pmatrix} \overline{I}_a \\ \overline{I}_b \end{pmatrix} = \begin{pmatrix} 10\angle 0^\circ \\ j\overline{I}_b - j\overline{I}_a + j\overline{I}_b \end{pmatrix}$$

reordering terms

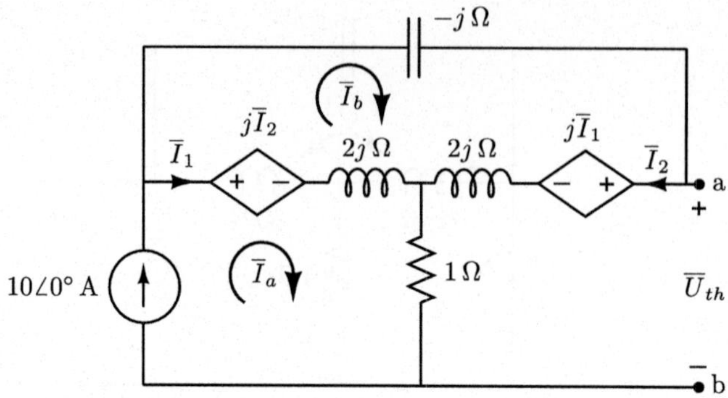

Fig. 2.108 Circuit for the problem resolution

$$\begin{pmatrix} 1 & 0 \\ -j & j \end{pmatrix} \begin{pmatrix} \overline{I}_a \\ \overline{I}_b \end{pmatrix} = \begin{pmatrix} 10\angle 0° \\ 0 \end{pmatrix}$$

and solving:

$$\overline{I}_a = \overline{I}_b = 10\angle 0° \text{ A}$$

Therefore:

$$\overline{I}_1 = \overline{I}_a - \overline{I}_b = 10\angle 0° - 10\angle 0° = 0 \text{ A}$$
$$\overline{I}_2 = \overline{I}_b = 10\angle 0° \text{ A}$$

Consequently, the Thévenin voltage is:

$$\overline{U}_{th} = j \cdot \overline{I}_1 + 2j \cdot \overline{I}_2 + 1 \cdot (\overline{I}_1 + \overline{I}_2) = 10 + 20j \text{ V}$$

Next, the Norton current is calculated by short-circuiting terminals a and b, as shown in Fig. 2.109. According to Fig. 2.109, applying mesh analysis:

$$\begin{pmatrix} 1 & 0 & 0 \\ -2j & 2j + 2j - j & -2j \\ -1 & -2j & 1 + 2j \end{pmatrix} \begin{pmatrix} \overline{I}_a \\ \overline{I}_b \\ \overline{I}_c \end{pmatrix} = \begin{pmatrix} 10\angle 0° \\ j\overline{I}_2 - j\overline{I}_1 \\ j\overline{I}_1 \end{pmatrix}$$

On the other hand:

$$\overline{I}_1 = \overline{I}_a - \overline{I}_b \ ; \ \overline{I}_2 = \overline{I}_b - \overline{I}_c$$

Substituting values

$$\begin{pmatrix} 1 & 0 & 0 \\ -2j & 2j + 2j - j & -2j \\ -1 & -2j & 1 + 2j \end{pmatrix} \begin{pmatrix} \overline{I}_a \\ \overline{I}_b \\ \overline{I}_c \end{pmatrix} = \begin{pmatrix} 10\angle 0° \\ j \cdot (\overline{I}_b - \overline{I}_c) - j \cdot (\overline{I}_a - \overline{I}_b) \\ j \cdot (\overline{I}_a - \overline{I}_b) \end{pmatrix}$$

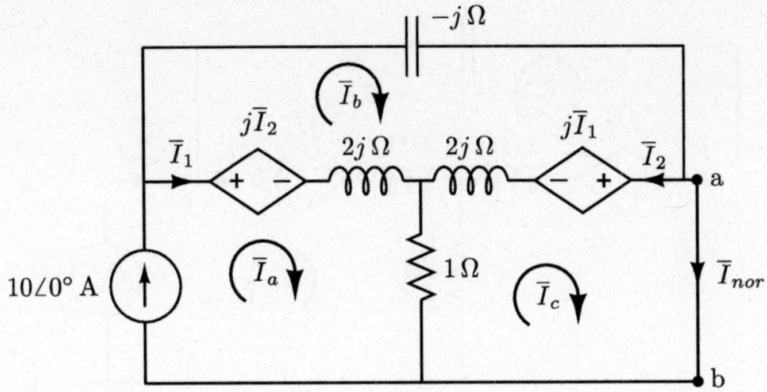

Fig. 2.109 Circuit with terminals a and b short-circuited

reordering terms

$$\begin{pmatrix} 1 & 0 & 0 \\ -j & j & -j \\ -1-j & -j & 1+2j \end{pmatrix} \begin{pmatrix} \overline{I}_a \\ \overline{I}_b \\ \overline{I}_c \end{pmatrix} = \begin{pmatrix} 10\angle 0° \\ 0 \\ 0 \end{pmatrix}$$

and solving yields the current $\overline{I}_c$, which corresponds to the Norton current, $\overline{I}_{nor}$:

$$\overline{I}_{nor} = \overline{I}_c = 15 + 5j \text{ A}$$

Once the Thévenin voltage and Norton current are obtained, the equivalent impedance between terminals a and b is calculated as follows:

$$\overline{Z}_{eq} = \frac{\overline{U}_{th}}{\overline{I}_{nor}} = \frac{10 + 20j}{15 + 5j} = 1 + j\,\Omega$$

Another way to calculate the equivalent impedance is from the passive circuit. In this case, since the circuit has dependent sources, the equivalent impedance cannot be obtained by associating impedances. Instead, a test source could be used connected between terminals a and b to the passive circuit as shown in Fig. 2.110. According to Fig. 2.110, applying mesh analysis:

$$\begin{pmatrix} 4j - j & -2j \\ -2j & 1+2j \end{pmatrix} \begin{pmatrix} \overline{I}_a \\ \overline{I}_b \end{pmatrix} = \begin{pmatrix} j\overline{I}_2 - j\overline{I}_1 \\ j\overline{I}_1 - \overline{U}_x \end{pmatrix}$$

On the other hand:

$$\overline{I}_1 = -\overline{I}_a \;\; ; \;\; \overline{I}_2 = \overline{I}_a - \overline{I}_b$$

Substituting values

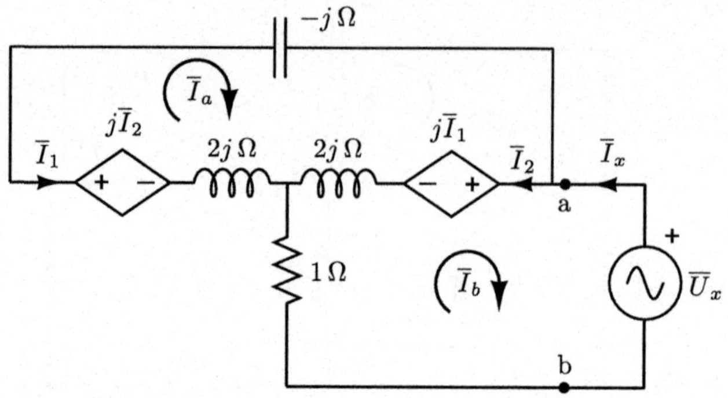

Fig. 2.110 Passive circuit

$$\begin{pmatrix} 4j - j & -2j \\ -2j & 1+2j \end{pmatrix} \begin{pmatrix} \overline{I}_a \\ \overline{I}_b \end{pmatrix} = \begin{pmatrix} j \cdot (\overline{I}_a - \overline{I}_b) + j\overline{I}_a \\ -j\overline{I}_a - \overline{U}_x \end{pmatrix} = \begin{pmatrix} 2j \cdot (\overline{I}_a - j\overline{I}_b \\ -j\overline{I}_a - \overline{U}_x \end{pmatrix}$$

reordering terms

$$\begin{pmatrix} j & -j \\ -j & 1+2j \end{pmatrix} \begin{pmatrix} \overline{I}_a \\ \overline{I}_b \end{pmatrix} = \begin{pmatrix} 0 \\ -\overline{U}_x \end{pmatrix}$$

and solving yields the current $\overline{I}_b$:

$$\overline{I}_b = \frac{-\overline{U}_x}{1+j}$$

Taking into account that $\overline{I}_b = -\overline{I}_x$, then the relationship between $\overline{U}_x$ and $\overline{I}_x$ is as follows:

$$\overline{U}_x = \overline{I}_x(1+j)$$

which allows the impedance to be determined as:

$$\overline{Z}_{eq} = \frac{\overline{U}_x}{\overline{I}_x} = \frac{1+j}{1+j} = 1 + j\,\Omega.$$

2.31 Thévenin and Norton Equivalents

In the circuit of Fig. 2.111, determine the Thévenin and Norton equivalents from terminals a and b.

Solution First, the coupled coils will be replaced by their representation with dependent voltage sources in the phasor domain as shown in Fig. 2.112.

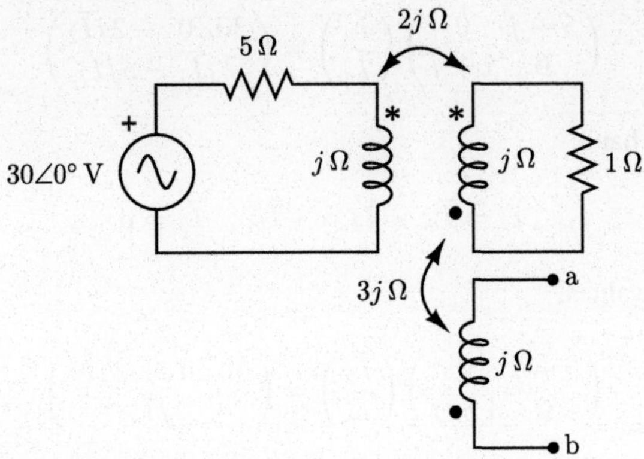

Fig. 2.111 Circuit considered in the problem under study

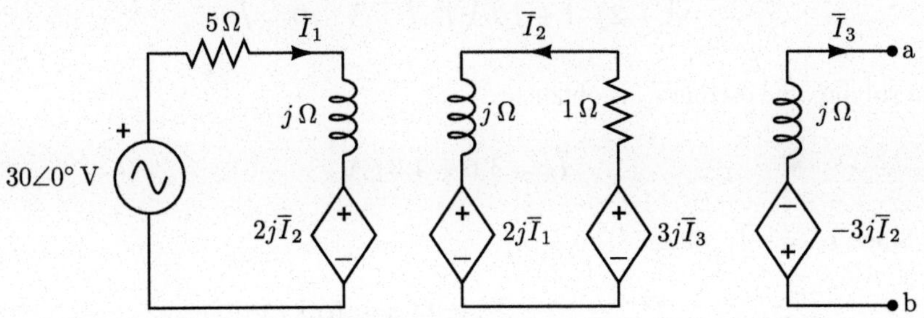

Fig. 2.112 Equivalent circuit for coupled coils

Next, the open circuit voltage between terminals a and b, Thévenin voltage, will be calculated. To do this, the circuit shown in Fig. 2.113 will be solved. According to Fig. 2.113:

$$\overline{U}_{th} = 3j \cdot \overline{I}_2$$

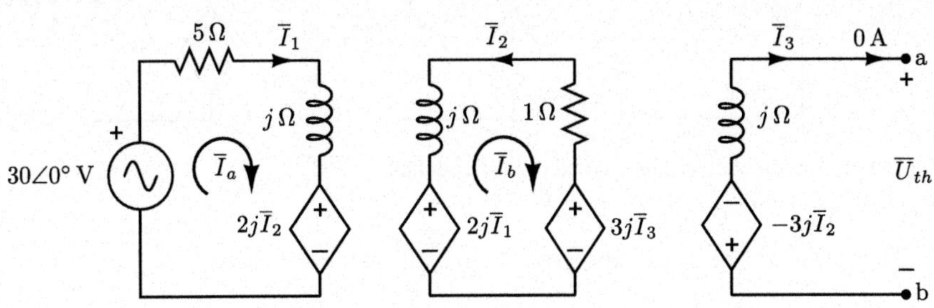

Fig. 2.113 Circuit considered in the problem under study

Applying the mesh analysis method:

$$\begin{pmatrix} 5+j & 0 \\ 0 & 1+j \end{pmatrix} \begin{pmatrix} \overline{I}_a \\ \overline{I}_b \end{pmatrix} = \begin{pmatrix} 30\angle 0^\circ - 2j\overline{I}_2 \\ 2j\overline{I}_1 - 3j\overline{I}_3 \end{pmatrix}$$

On the other hand:

$$\overline{I}_1 = \overline{I}_a \ ; \ \overline{I}_2 = -\overline{I}_b \ ; \ \overline{I}_3 = 0$$

Substituting values:

$$\begin{pmatrix} 5+j & 0 \\ 0 & 1+j \end{pmatrix} \begin{pmatrix} \overline{I}_a \\ \overline{I}_b \end{pmatrix} = \begin{pmatrix} 30\angle 0^\circ + 2j\overline{I}_b \\ 2j\overline{I}_a \end{pmatrix}$$

rearranging terms:

$$\begin{pmatrix} 5+j & -2j \\ -2j & 1+j \end{pmatrix} \begin{pmatrix} \overline{I}_a \\ \overline{I}_b \end{pmatrix} = \begin{pmatrix} 30\angle 0^\circ \\ 0 \end{pmatrix}$$

and solving, the current $\overline{I}_b$ is obtained:

$$\overline{I}_b = 3{,}6 + 4{,}8j \text{ A}$$

Since $\overline{I}_2 = -\overline{I}_b$, then:

$$\overline{U}_{th} = -3j \cdot \overline{I}_b = 14{,}4 - 10{,}8j \text{ V}$$

Next, the Norton current is calculated, which is obtained by short-circuiting terminals a and b, as shown in Fig. 2.114.

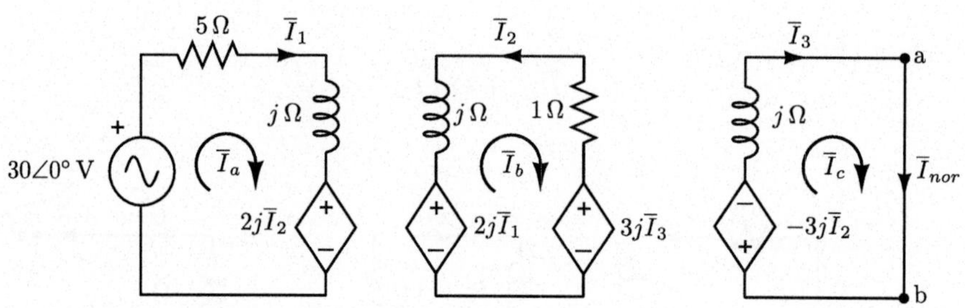

Fig. 2.114 Circuit with terminals a and b short-circuited

According to Fig. 2.114, applying mesh analysis:

$$\begin{pmatrix} 5+j & 0 & 0 \\ 0 & 1+j & 0 \\ 0 & 0 & j \end{pmatrix} \begin{pmatrix} \overline{I}_a \\ \overline{I}_b \\ \overline{I}_c \end{pmatrix} = \begin{pmatrix} 30\angle 0^\circ - 2j\overline{I}_2 \\ 2j\overline{I}_1 - 3j\overline{I}_3 \\ 3j\overline{I}_2 \end{pmatrix}$$

On the other hand:

$$\overline{I}_1 = \overline{I}_a \ ; \ \overline{I}_2 = -\overline{I}_b \ ; \ \overline{I}_3 = \overline{I}_c$$

Substituting values:

$$\begin{pmatrix} 5+j & 0 & 0 \\ 0 & 1+j & 0 \\ 0 & 0 & j \end{pmatrix} \begin{pmatrix} \overline{I}_a \\ \overline{I}_b \\ \overline{I}_c \end{pmatrix} = \begin{pmatrix} 30\angle 0^\circ + 2j\overline{I}_b \\ 2j\overline{I}_a - 3j\overline{I}_c \\ -3j\overline{I}_b \end{pmatrix}$$

rearranging terms:

$$\begin{pmatrix} 5+j & -2j & 0 \\ -2j & 1+j & 3j \\ 0 & 3j & j \end{pmatrix} \begin{pmatrix} \overline{I}_a \\ \overline{I}_b \\ \overline{I}_c \end{pmatrix} = \begin{pmatrix} 30\angle 0^\circ \\ 0 \\ 0 \end{pmatrix}$$

and solving, the current $\overline{I}_c$, which corresponds to the Norton current $\overline{I}_{nor}$, is obtained:

$$\overline{I}_{nor} = \overline{I}_c \approx 3{,}88 - 1{,}69j \text{ A}$$

Once the Thévenin voltage and Norton current are obtained, the equivalent impedance between terminals a and b is calculated as follows:

$$\overline{Z}_{eq} = \frac{\overline{U}_{th}}{\overline{I}_{nor}} = \frac{14{,}4 - 10{,}8j}{3{,}88 - 1{,}69j} \approx 4{,}14 - 0{,}98j \ \Omega$$

Another way to calculate the equivalent impedance is from the passive circuit. In this case, since the circuit has dependent sources, the equivalent impedance cannot be obtained by associating impedances. Instead, a test source connected between terminals a and b to the passive circuit could be used, as shown in Fig. 2.115.

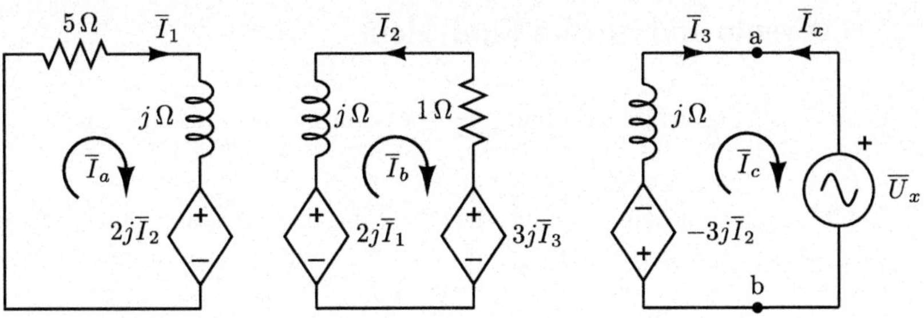

Fig. 2.115 Passive circuit

According to Fig. 2.115, applying mesh analysis:

$$\begin{pmatrix} 5+j & 0 & 0 \\ 0 & 1+j & 0 \\ 0 & 0 & j \end{pmatrix} \begin{pmatrix} \overline{I}_a \\ \overline{I}_b \\ \overline{I}_c \end{pmatrix} = \begin{pmatrix} -2j\overline{I}_2 \\ 2j\overline{I}_1 - 3j\overline{I}_3 \\ 3j\overline{I}_2 - \overline{U}_x \end{pmatrix}$$

On the other hand:

$$\overline{I}_1 = \overline{I}_a \; ; \; \overline{I}_2 = -\overline{I}_b \; ; \; \overline{I}_3 = \overline{I}_c$$

Substituting values:

$$\begin{pmatrix} 5+j & 0 & 0 \\ 0 & 1+j & 0 \\ 0 & 0 & j \end{pmatrix} \begin{pmatrix} \overline{I}_a \\ \overline{I}_b \\ \overline{I}_c \end{pmatrix} = \begin{pmatrix} 2j\overline{I}_b \\ 2j\overline{I}_a - 3j\overline{I}_c \\ -3j\overline{I}_b - \overline{U}_x \end{pmatrix}$$

rearranging terms:

$$\begin{pmatrix} 5+j & -2j & 0 \\ -2j & 1+j & 3j \\ 0 & 3j & j \end{pmatrix} \begin{pmatrix} \overline{I}_a \\ \overline{I}_b \\ \overline{I}_c \end{pmatrix} = \begin{pmatrix} 0 \\ 0 \\ -\overline{U}_x \end{pmatrix}$$

and solving, the current $\overline{I}_c$ is obtained:

$$\overline{I}_c \approx \frac{\overline{U}_x}{-4{,}14 + 0{,}98j}$$

Considering that $\overline{I}_c = -\overline{I}_x$, then the relationship between $\overline{U}_x$ and $\overline{I}_x$ is as follows:

$$\overline{Z}_{eq} = \frac{\overline{U}_x}{\overline{I}_x} = 4{,}14 - 0{,}98j\ \Omega.$$

2.32 Thévenin and Norton Equivalents

In the circuit of Fig. 2.116, calculate the Thévenin and Norton equivalents from terminals a and b.

Solution First, the Thévenin voltage, which is the open circuit voltage between terminals a and b, will be obtained. To do this, the circuit shown in Fig. 2.117 will be solved.

According to Fig. 2.117:

$$\overline{U} = \overline{U}_{th} = 0{,}5\overline{I} \cdot 1 \;\Rightarrow\; \overline{I} = \frac{\overline{U}_{th}}{0{,}5} \tag{2.4}$$

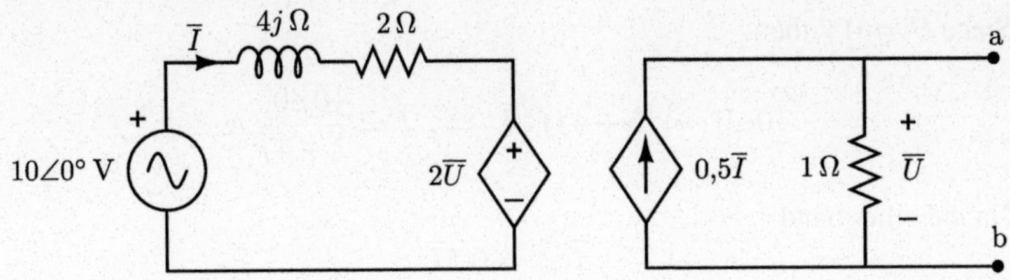

Fig. 2.116 Circuit considered in the problem under study

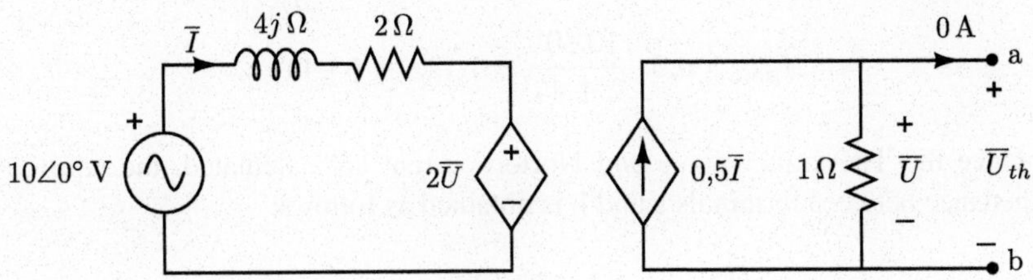

Fig. 2.117 Circuit considered in the problem under study

On the other hand:

$$10\angle 0^\circ = (2+4j)\cdot \overline{I} + 2\overline{U} \tag{2.5}$$

Substituting equation (2.4) into Eq. (2.5) results in:

$$10\angle 0^\circ = (2+4j)\cdot \frac{\overline{U}_{th}}{0{,}5} + 2\overline{U}_{th} \Rightarrow \overline{U}_{th} = \frac{10\angle 0^\circ}{2+\dfrac{2+4j}{0{,}5}} \approx 1\angle -53{,}13^\circ \text{ V}$$

Next, the Norton current, which is obtained by short-circuiting terminals a and b, will be calculated, as shown in Fig. 2.118. According to Fig. 2.118:

$$10\angle 0^\circ = (2+4j)\cdot \overline{I} + 2\overline{U}$$

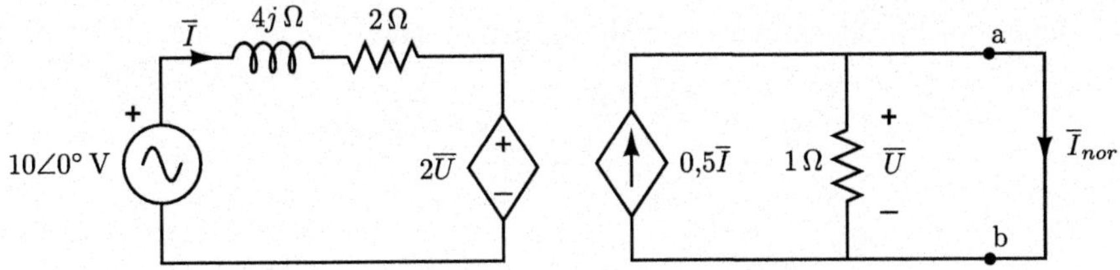

Fig. 2.118 Circuit with terminals a and b short-circuited

Since $\overline{U} = 0\,\text{V}$ then:

$$10\angle 0^\circ = (2+4j)\cdot \overline{I} \;\Rightarrow\; \overline{I} = \frac{10\angle 0^\circ}{2+4j}\,\text{A}$$

On the other hand:

$$\overline{I}_{nor} = 0{,}5\overline{I}$$

Substituting values results in:

$$\overline{I}_{nor} = 0{,}5\cdot\frac{10\angle 0^\circ}{2+4j} \approx 1{,}12\angle -63{,}43^\circ\,\text{A}$$

Once the Thévenin voltage and Norton current are calculated, the equivalent impedance between terminals a and b is obtained as follows:

$$\overline{Z}_{eq} = \frac{\overline{U}_{th}}{\overline{I}_{nor}} = \frac{1\angle -53{,}13^\circ}{1{,}12\angle -63{,}43^\circ} \approx 0{,}88 + 0{,}16j\,\Omega$$

An alternative way to obtain the equivalent impedance is by connecting a test source between a and b in the passive circuit, as shown in Fig. 2.119.

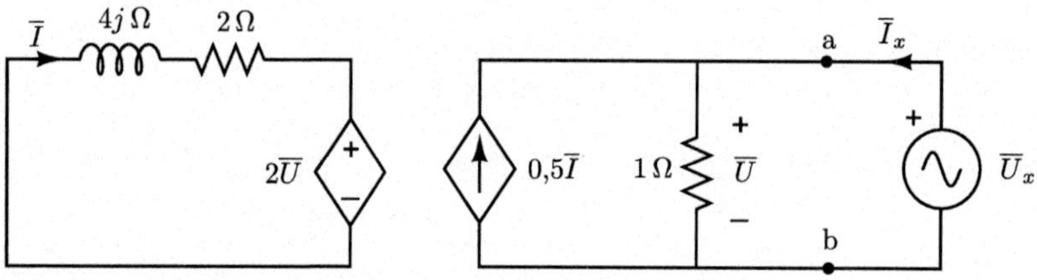

Fig. 2.119 Circuit with a test voltage source

According to Fig. 2.119, applying Kirchhoff's voltage law:

$$(2+4j)\overline{I} + 2\overline{U} = 0$$

On the other hand, applying Kirchhoff's current law:

$$0{,}5\overline{I} + \overline{I}_x = \frac{\overline{U}}{1} \;\Rightarrow\; \overline{I} = \frac{\overline{U} - \overline{I}_x}{0{,}5}$$

Considering that $\overline{U} = \overline{U}_x$ then:

$$(2+4j)\cdot\left(\frac{\overline{U}_x - \overline{I}_x}{0{,}5}\right) + 2\overline{U}_x = 0 \;\Rightarrow\; \overline{Z}_{eq} = \frac{\overline{U}_x}{\overline{I}_x} = \frac{4+8j}{6+8j} = 0{,}88 + 0{,}16j\,\Omega.$$

2.33 Thévenin and Norton Equivalents

In the circuit of Fig. 2.120, calculate the Thévenin and Norton equivalents from terminals a and b.

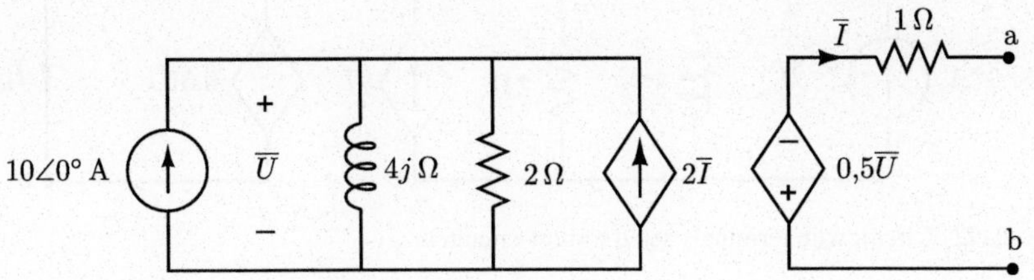

Fig. 2.120 Circuit considered in the problem under study

Solution First, the open-circuit voltage between a and b, Thévenin voltage, will be obtained. To do this, the circuit shown in Fig. 2.121 will be solved.

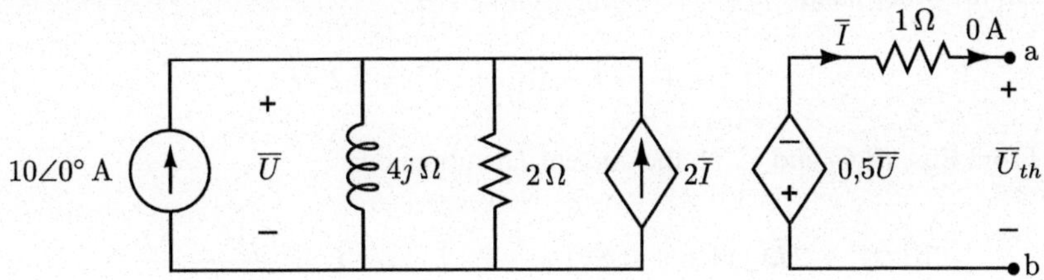

Fig. 2.121 Circuit for the calculation of Thévenin voltage

According to Fig. 2.121, applying the nodal analysis method on the left side of the circuit yields:

$$\left(\frac{1}{4j}+\frac{1}{2}\right)\cdot\overline{U}=10\angle 0^{\circ}+2\overline{I}\ \Rightarrow\ \overline{U}=(10\angle 0^{\circ}+2\overline{I})\cdot(1{,}6+0{,}8j)$$

Since $\overline{I}=0$, then:

$$\overline{U}=10\angle 0^{\circ}\cdot(1{,}6+0{,}8j)=16+8j\ \text{V}$$

Consequently, the Thévenin voltage is as follows:

$$\overline{U}_{th}=-0{,}5\overline{U}=-0{,}5\cdot(16+8j)=-8-4j\ \text{V}$$

Next, the Norton current is calculated, which is obtained by short-circuiting terminals a and b, as shown in Fig. 2.122.

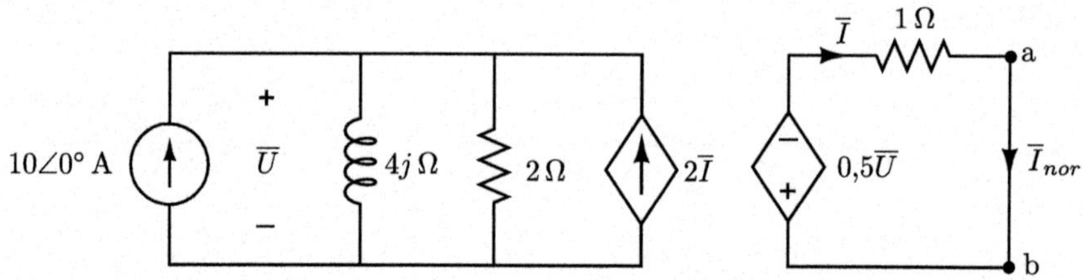

Fig. 2.122 Circuit with terminals a and b short-circuited

According to Fig. 2.122, applying the nodal analysis method again on the left side of the circuit yields:

$$\left(\frac{1}{4j}+\frac{1}{2}\right)\cdot\overline{U}=10\angle 0°+2\overline{I}\ \Rightarrow\ \overline{U}=(10\angle 0°+2\overline{I})\cdot(1{,}6+0{,}8j) \qquad (2.6)$$

On the other hand:

$$0{,}5\overline{U}+1\cdot\overline{I}=0\ \Rightarrow\ \overline{U}=-2\cdot\overline{I} \qquad (2.7)$$

From Eqs. (2.6) and (2.7), the current $\overline{I}$ is obtained:

$$(10\angle 0°+2\overline{I})\cdot(1{,}6+0{,}8j)=-2\cdot\overline{I}\ \Rightarrow\ \overline{I}=\frac{-16-8j}{5{,}2+1{,}6j}\ \text{A}$$

Since $\overline{I}=\overline{I}_{nor}$, then:

$$\overline{I}_{nor}=\frac{-16-8j}{5{,}2+1{,}6j}\ \text{A}$$

Once the Thévenin voltage and Norton current are calculated, the equivalent impedance between terminals a and b is obtained as follows:

$$\overline{Z}_{eq}=\frac{\overline{U}_{th}}{\overline{I}_{nor}}=\frac{-8-4j}{\dfrac{-16-8j}{5{,}2+1{,}6j}}=2{,}6+0{,}8j\ \Omega$$

An alternative way to obtain the equivalent impedance is by connecting a test source between a and b to the passive circuit, as shown in Fig. 2.123.

According to Fig. 2.123, applying the nodal analysis method on the left side of the circuit yields:

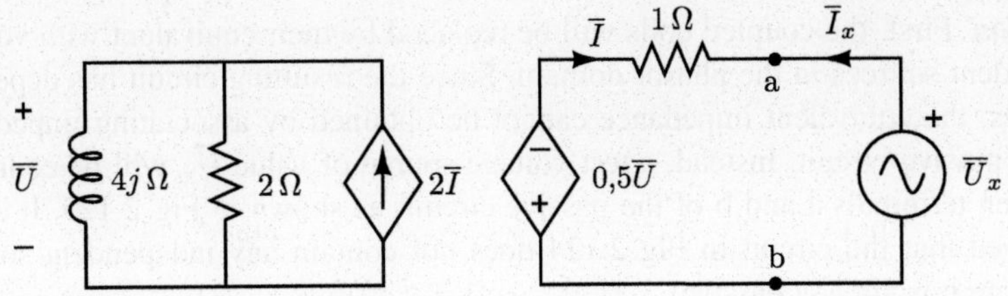

Fig. 2.123 Circuit with a test voltage source

$$\left(\frac{1}{4j}+\frac{1}{2}\right)\cdot\overline{U}=2\overline{I}\ \Rightarrow\ \overline{U}=2\overline{I}\cdot(1{,}6+0{,}8j) \tag{2.8}$$

On the other hand:

$$0,5\overline{U}+1\cdot\overline{I}+\overline{U}_x=0\ \Rightarrow\ \overline{U}_x=-1\cdot\overline{I}-0{,}5\overline{U} \tag{2.9}$$

Substituting equation (2.8) into Eq. (2.9):

$$\overline{U}_x=-1\cdot\overline{I}-0{,}5\cdot 2\overline{I}\cdot(1{,}6+0{,}8j)=\overline{I}\cdot(-2{,}6-0{,}8j)$$

Considering that $\overline{I}=-\overline{I}_x$, then:

$$\overline{Z}_{eq}=\frac{\overline{U}_x}{\overline{I}_x}=2{,}6+0{,}8j\ \Omega.$$

2.34 Equivalent Impedance

In the circuit of Fig. 2.124, obtain the equivalent impedance between terminals a and b.

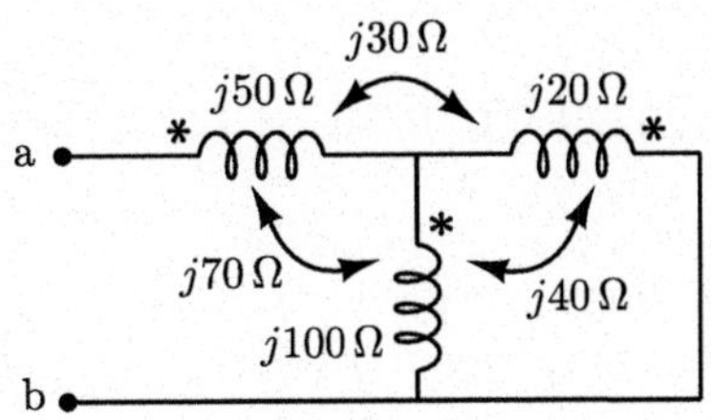

Fig. 2.124 Circuit considered in the problem under study

Solution First, the coupled coils will be replaced by their equivalent with voltage-dependent sources in the phasor domain. Since the resulting circuit has dependent sources, the equivalent impedance cannot be obtained by associating impedances in the passive circuit. Instead, a test voltage source of value $\overline{U}_x$ will be connected between terminals a and b of the passive circuit, as shown in Fig. 2.125. It can be observed that the circuit in Fig. 2.124 does not contain any independent sources, therefore it is already passive.

According to Fig. 2.125, applying the mesh method:

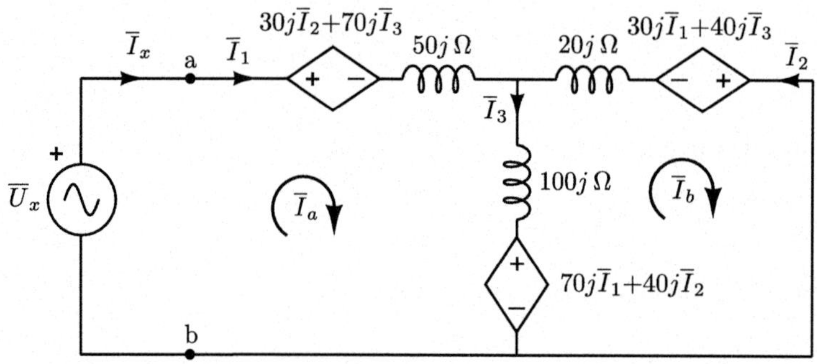

Fig. 2.125 Equivalent model of coupled coils

$$\begin{pmatrix} 150j & -100j \\ -100j & 120j \end{pmatrix} \begin{pmatrix} \overline{I}_a \\ \overline{I}_b \end{pmatrix} = \begin{pmatrix} \overline{U}_x - 30j\overline{I}_2 - 70j\overline{I}_3 - 70j\overline{I}_1 - 40j\overline{I}_2 \\ 70j\overline{I}_1 + 40j\overline{I}_2 + 30j\overline{I}_1 + 40j\overline{I}_3 \end{pmatrix}$$

On the other hand:

$$\overline{I}_1 = \overline{I}_a \ ; \ \overline{I}_2 = -\overline{I}_b \ ; \ \overline{I}_3 = \overline{I}_a - \overline{I}_b$$

Substituting values

$$\begin{pmatrix} 150j & -100j \\ -100j & 120j \end{pmatrix} \begin{pmatrix} \overline{I}_a \\ \overline{I}_b \end{pmatrix} = \begin{pmatrix} \overline{U}_x - 140j\overline{I}_a + 140j\overline{I}_b \\ 140j\overline{I}_a - 80j\overline{I}_b \end{pmatrix}$$

reordering terms

$$\begin{pmatrix} 290j & -240j \\ -240j & 200j \end{pmatrix} \begin{pmatrix} \overline{I}_a \\ \overline{I}_b \end{pmatrix} = \begin{pmatrix} \overline{U}_x \\ 0 \end{pmatrix}$$

and solving yields the current $\overline{I}_a$:

$$\overline{I}_a = \frac{\overline{U}_x}{290j - \dfrac{240 \cdot 240j}{200}}$$

Considering that $\overline{I}_a = \overline{I}_x$, the relationship between $\overline{U}_x$ and $\overline{I}_x$ is as follows:

$$\overline{Z}_{eq} = \frac{\overline{U}_x}{\overline{I}_x} = 290j - \frac{240 \cdot 240j}{200} = 2j\,\Omega.$$

2.35 Equivalent Impedance

In the circuit of Fig. 2.126, it is known that the equivalent impedance between terminals a and b has only a resistive part. Determine the value of X_M.

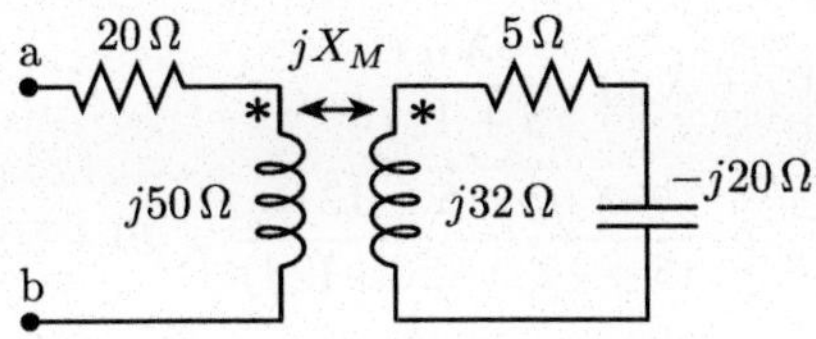

Fig. 2.126 Circuit considered in the problem under study

Solution First, the coupled coils will be replaced by their representation with voltage-dependent sources in the phasor domain. Since the resulting circuit has dependent sources, the equivalent impedance cannot be obtained by associating impedances in the passive circuit. Instead, a test voltage source of value $\overline{U}_x$ will be connected between terminals a and b of the passive circuit, as shown in Fig. 2.127. It can be observed that the circuit in Fig. 2.126 does not contain any independent sources, therefore it is already passive.

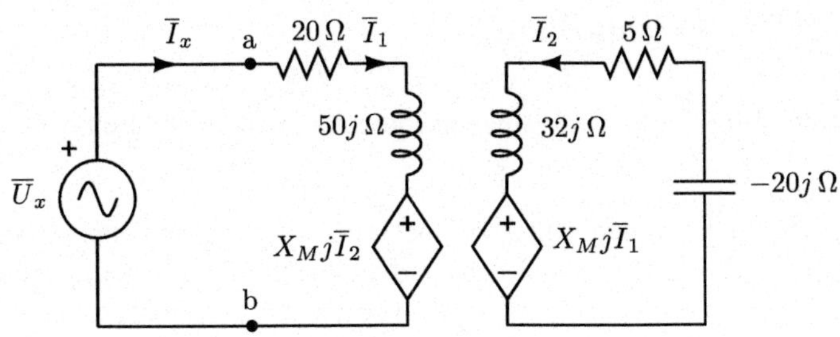

Fig. 2.127 Equivalent model of coupled coils

According to Fig. 2.127, applying Kirchhoff's voltage law in the left part of the circuit yields:

$$(20+50j)\cdot\overline{I}_1+X_M j\overline{I}_2-\overline{U}_x=0 \;\Rightarrow\; \overline{U}_x=(20+50j)\cdot\overline{I}_1+X_M j\overline{I}_2$$

Applying Kirchhoff's voltage law in the right part of the circuit yields:

$$(5-20j+32j)\cdot\overline{I}_2+X_M j\overline{I}_1=0 \;\Rightarrow\; \overline{I}_2=\frac{-X_M j}{5+12j}\cdot\overline{I}_1$$

Since $\overline{I}_1=\overline{I}_x$, then:

$$\overline{U}_x=(20+50j)\cdot\overline{I}_1+X_M j\cdot\overline{I}_2=\left[20+50j+X_M j\cdot\frac{-X_M j}{5+12j}\right]\cdot\overline{I}_x$$

Therefore, the equivalent impedance is:

$$\begin{aligned}\overline{Z}_{eq}&=\frac{\overline{U}_x}{\overline{I}_x}=20+50j+X_M j\cdot\frac{-X_M j}{5+12j}\\&=20+50j+X_M j\cdot\frac{-X_M j\cdot(5-12j)}{(5+12j)\cdot(5-12j)}=20+50j+\frac{X_M^2\cdot(5-12j)}{5^2+12^2}\\&=\underbrace{\left[20+\frac{5X_M^2}{5^2+12^2}\right]}_{R_{eq}}+\underbrace{\left[50-\frac{12X_M^2}{5^2+12^2}\right]}_{X_{eq}}j\end{aligned}$$

For the equivalent impedance to be purely resistive, the imaginary part must be zero:

$$X_{eq}=0 \;\Rightarrow\; 50-\frac{12X_M^2}{5^2+12^2}=0 \;\Rightarrow X_M\approx 26{,}54\,\Omega$$

The solution $X_M=-26{,}54\,\Omega$ has been discarded since X_M must be positive.

2.36 Equivalent Impedance

In the circuit of Fig. 2.128, determine the equivalent impedance seen from terminals a and b.

Solution First, the coupled coils will be replaced by their representation with voltage-dependent sources in the phasor domain. Since the resulting circuit has dependent sources, the equivalent impedance cannot be obtained by associating impedances in the passive circuit. Instead, a test voltage source of value $\overline{U}_x$ will be

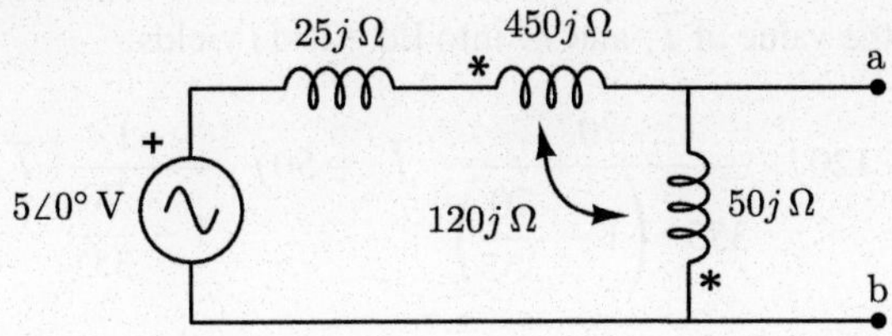

Fig. 2.128 Circuit considered in the problem under study

connected between terminals a and b of the passive circuit, as shown in Fig. 2.129. Applying Kirchhoff's voltage law in the left part of the circuit yields:

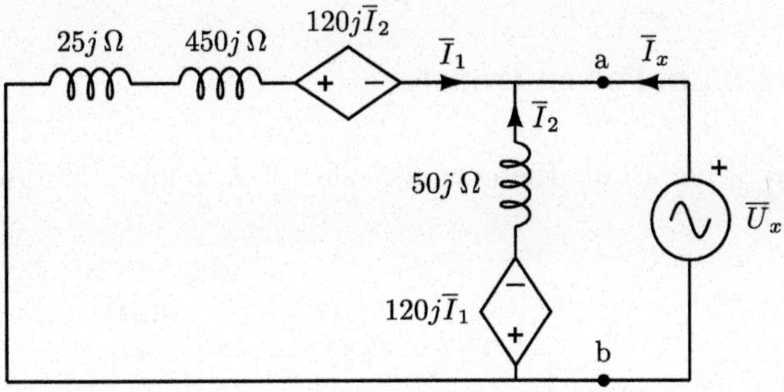

Fig. 2.129 Equivalent model of coupled coils

$$(25j + 450j) \cdot \overline{I}_1 + 120j \cdot \overline{I}_2 - 50j \cdot \overline{I}_2 - 120j \cdot \overline{I}_1 = 0 \;\Rightarrow\; 355j \cdot \overline{I}_1 + 70j \cdot \overline{I}_2 = 0 \tag{2.10}$$

Applying Kirchhoff's voltage law in the right part of the circuit yields:

$$\overline{U}_x + 120j \cdot \overline{I}_1 + 50j \cdot \overline{I}_2 = 0 \tag{2.11}$$

Finally, taking into account Kirchhoff's current law:

$$\overline{I}_1 + \overline{I}_2 + \overline{I}_x = 0 \tag{2.12}$$

Solving Eqs. (2.10) and (2.12) yields the currents $\overline{I}_1$ and $\overline{I}_2$ in terms of $\overline{I}_x$:

$$\overline{I}_1 = \frac{70}{355 \cdot \left(1 - \dfrac{70}{355}\right)} \cdot \overline{I}_x \;;\; \overline{I}_2 = \frac{-1}{1 - \dfrac{70}{355}} \cdot \overline{I}_x$$

Substituting the value of $\overline{I}_1$ and $\overline{I}_2$ into Eq. (2.11) yields:

$$\overline{U}_x + 120j \cdot \frac{70}{355 \cdot \left(1 - \frac{70}{355}\right)} \cdot \overline{I}_x + 50j \cdot \frac{-1}{1 - \frac{70}{355}} \cdot \overline{I}_x = 0$$

Therefore, the equivalent impedance seen from terminals a and b is:

$$\overline{Z}_{eq} = \frac{\overline{U}_x}{\overline{I}_x} = \frac{-120j \cdot 70}{355 \cdot \left(1 - \frac{70}{355}\right)} + \frac{50j}{1 - \frac{70}{355}} \approx 32{,}81j\,\Omega.$$

2.37 Maximum Power Transfer

In the circuit of Fig. 2.130, determine the value of $\overline{Z}$ so that it absorbs maximum power.

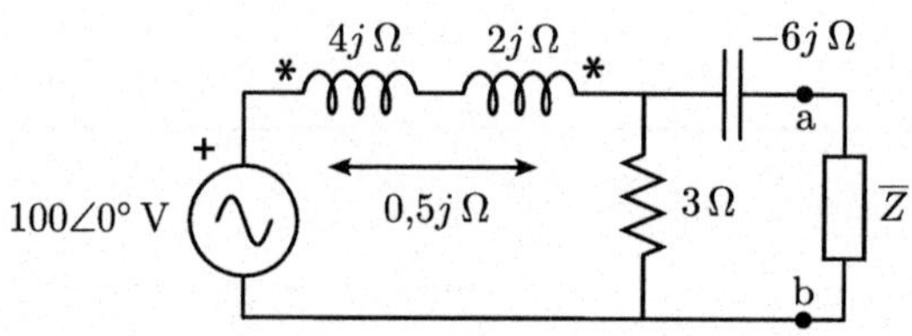

Fig. 2.130 Circuit considered in the problem under study

Solution First, the coupled coils will be replaced by their representation with voltage-dependent sources in the phasor domain, as shown in Fig. 2.131.

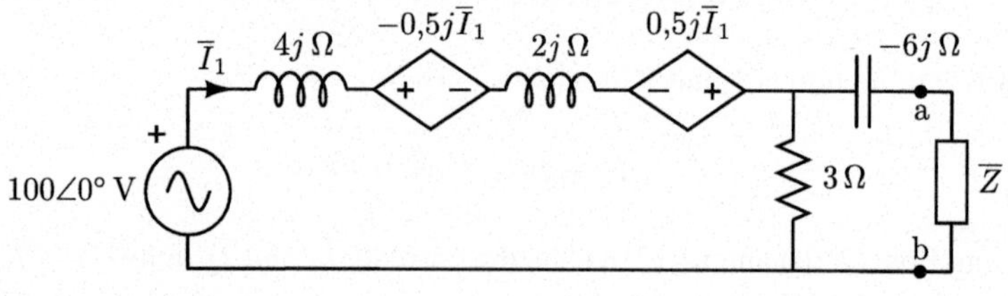

Fig. 2.131 Equivalent model of coupled coils

The circuit of Fig. 2.131 can be simplified by associating the two coils and the two dependent sources, resulting in the circuit of Fig. 2.132.

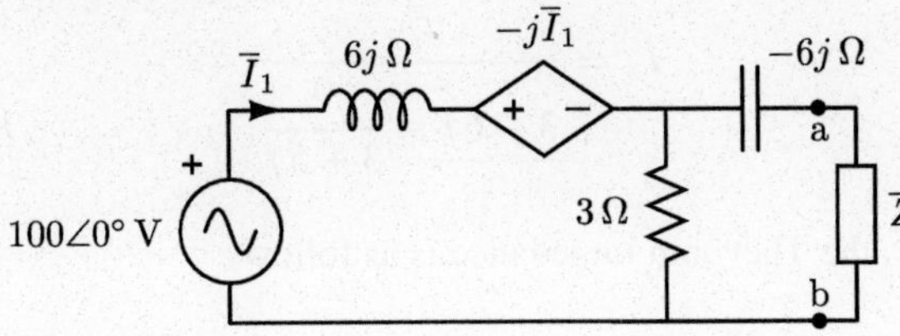

Fig. 2.132 Simplified circuit

Next, the equivalent impedance from terminals a and b, Thévenin impedance, will be calculated. For this, the independent sources of the circuit will be zeroed out and a test source will be connected between terminals a and b as shown in Fig. 2.133, where two mesh currents have also been represented for their resolution.

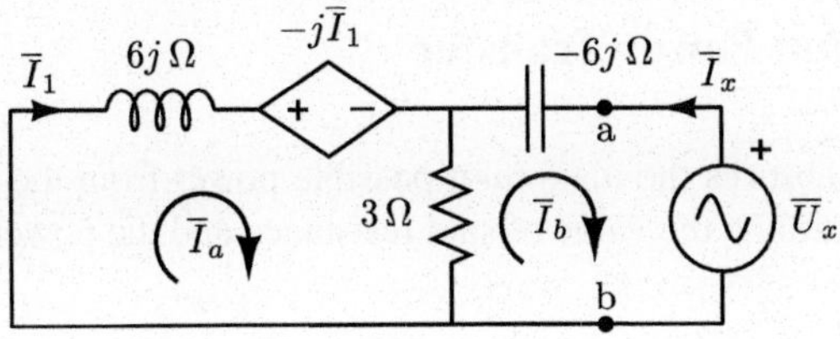

Fig. 2.133 Circuit with a test voltage source

According to Fig. 2.133, applying the mesh analysis method:

$$\begin{pmatrix} 3+6j & -3 \\ -3 & 3-6j \end{pmatrix} \begin{pmatrix} \overline{I}_a \\ \overline{I}_b \end{pmatrix} = \begin{pmatrix} j \cdot \overline{I}_1 \\ -\overline{U}_x \end{pmatrix}$$

On the other hand:

$$\overline{I}_1 = \overline{I}_a \ ; \ \overline{I}_b = -\overline{I}_x$$

Substituting values:

$$\begin{pmatrix} 3+6j & -3 \\ -3 & 3-6j \end{pmatrix} \begin{pmatrix} \overline{I}_a \\ -\overline{I}_x \end{pmatrix} = \begin{pmatrix} j \cdot \overline{I}_a \\ -\overline{U}_x \end{pmatrix}$$

rearranging terms:

$$\begin{pmatrix} 3+5j & -3 \\ -3 & 3-6j \end{pmatrix} \begin{pmatrix} \overline{I}_a \\ -\overline{I}_x \end{pmatrix} = \begin{pmatrix} 0 \\ -\overline{U}_x \end{pmatrix}$$

and solving, the current $\overline{I}_x$ is obtained:

$$\overline{I}_x = \frac{\overline{U}_x}{3 - 6j - \dfrac{9}{3+5j}}$$

Consequently, the Thévenin impedance is as follows:

$$\overline{Z}_{th} = \frac{\overline{U}_x}{\overline{I}_x} = 3 - 6j - \frac{9}{3+5j} \approx 2{,}2 - 4{,}67j\ \Omega$$

Once the Thévenin impedance from terminals a and b is calculated, the impedance $\overline{Z}$ that, once connected between these terminals, absorbs the maximum power is as follows:

$$\overline{Z} = \overline{Z}_{th}^{*} = 2{,}2 + 4{,}67j\ \Omega.$$

2.38 Maximum Power Transfer

The resistance R absorbs the maximum possible power from the rest of the circuit in Fig. 2.134. Determine the value of said resistance and the power it absorbs.

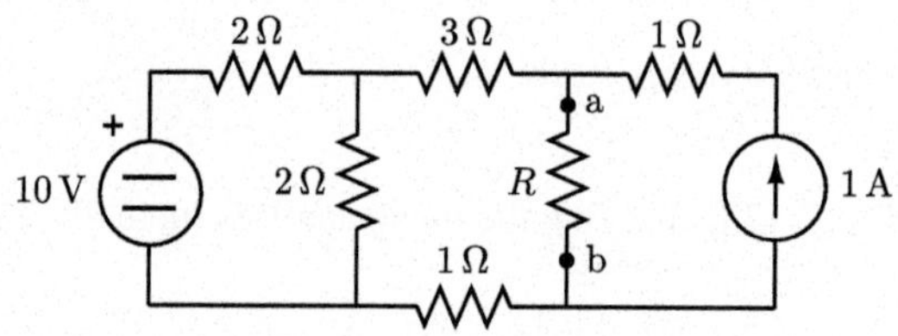

Fig. 2.134 Circuit considered in the problem under study

Solution To calculate the resistance R that absorbs the maximum power and the value of said power, it is necessary to obtain the Thévenin equivalent from terminals a and b.

First, the open-circuit voltage between terminals a and b, Thévenin voltage, will be calculated. For this, the circuit shown in Fig. 2.135 will be solved.

The circuit in Fig. 2.135 can be solved using superposition, resulting in the two circuits in Fig. 2.136.

According to circuit 1 in Fig. 2.136, the voltage U_{th1} is obtained by applying the voltage divider concept:

$$U_{th1} = 10 \cdot \frac{2}{2+2} = 5\ \text{V}$$

On the other hand, circuit 2 in Fig. 2.136 can be simplified, as shown in Fig. 2.137.

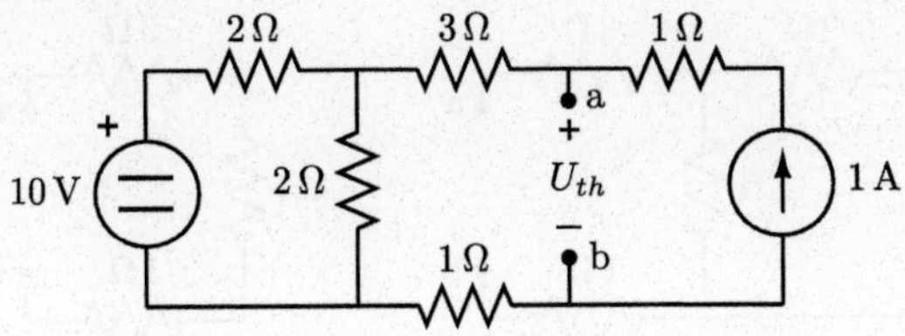

Fig. 2.135 Circuit considered in the problem under study

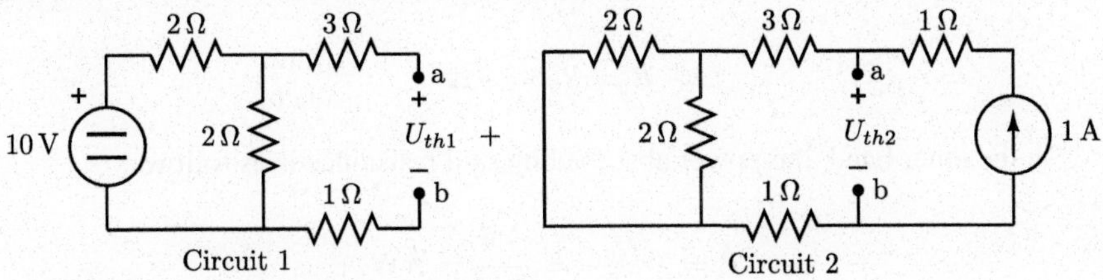

Fig. 2.136 Circuits after applying superposition

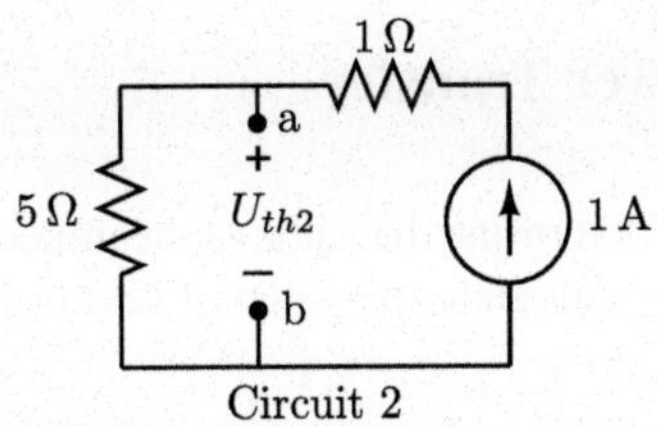

Fig. 2.137 Simplification of circuit 2

According to Fig. 2.137:

$$U_{th2} = 5 \cdot 1 = 5\,\text{V}$$

Once the voltages U_{th1} and U_{th2} are calculated, the Thévenin voltage U_{th} is as follows:

$$U_{th} = U_{th1} + U_{th2} = 5 + 5 = 10\,\text{V}$$

Next, the equivalent resistance, Thévenin resistance, from terminals a and b will be calculated. For this, the sources of the circuit in Fig. 2.134 will be zeroed out, as shown in Fig. 2.138.

According to Fig. 2.138, the Thévenin resistance is as follows:

$$R_{th} = 3 + 1 + 1 = 5\,\Omega$$

Once the equivalent resistance from terminals a and b is calculated, the resistance R that absorbs the maximum power is as follows:

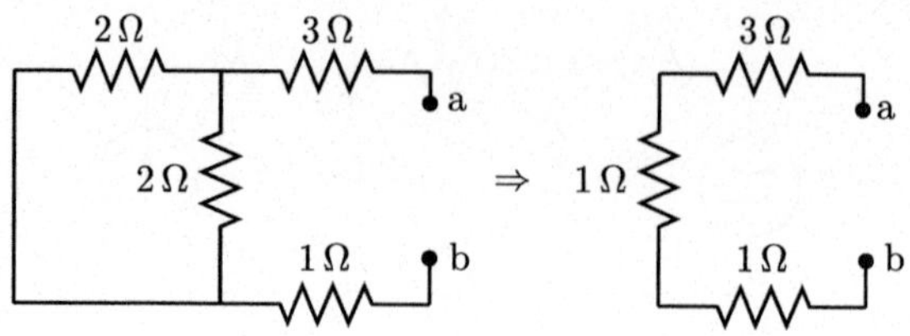

Fig. 2.138 Passive circuit

$$R = R_{th} = 5\,\Omega$$

On the other hand, the power absorbed by said resistance is as follows:

$$P = \frac{U_{th}^2}{4 \cdot R_{th}} = \frac{10^2}{4 \cdot 5} = 5\,\text{W}.$$

2.39 Maximum Power Transfer

In the circuit of Fig. 2.139, determine the value of the impedance $\overline{Z}_L$ so that it receives the maximum power. Also, calculate the value of said power.

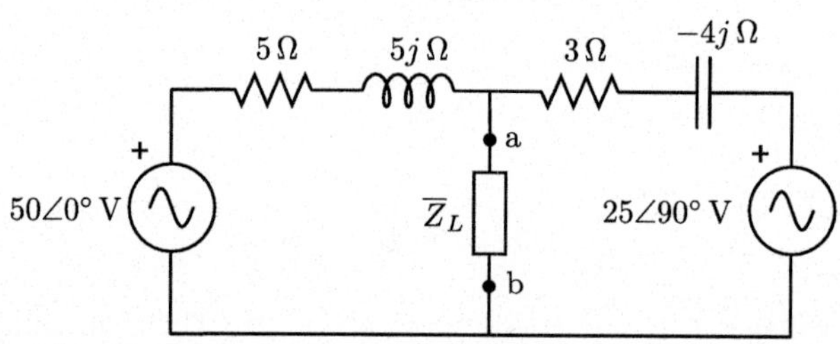

Fig. 2.139 Circuit considered in the problem under study

Solution To calculate the impedance $\overline{Z}_L$ that absorbs the maximum power and the value of said power, it is necessary to obtain the Thévenin equivalent from terminals a and b.

First, the open-circuit voltage between terminals a and b, Thévenin voltage, will be calculated. For this, the circuit shown in Fig. 2.140 will be solved. It can be noted that the impedance $\overline{Z}_L$ has not been considered since it is not part of the equivalent.

According to Fig. 2.140:

$$\overline{I} = \frac{50\angle 0^\circ - 25\angle 90^\circ}{5 + 5j + 3 - 4j} \approx 6{,}93\angle - 33{,}7^\circ\ \text{A}$$

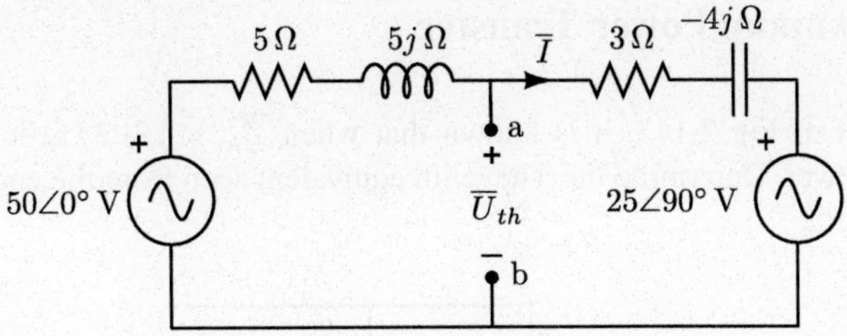

Fig. 2.140 Circuit considered in the problem under study

Therefore:

$$\overline{U}_{th} = (3-4j)\cdot\overline{I} + 25\angle 90^\circ = (3-4j)\cdot 6{,}93\angle -33{,}7^\circ + 25\angle 90^\circ \approx 9{,}81\angle -78{,}69^\circ \text{ V}$$

Next, the equivalent impedance, Thévenin impedance, from terminals a and b is calculated. For this, the circuit in Fig. 2.139 is made passive, as shown in Fig. 2.141.

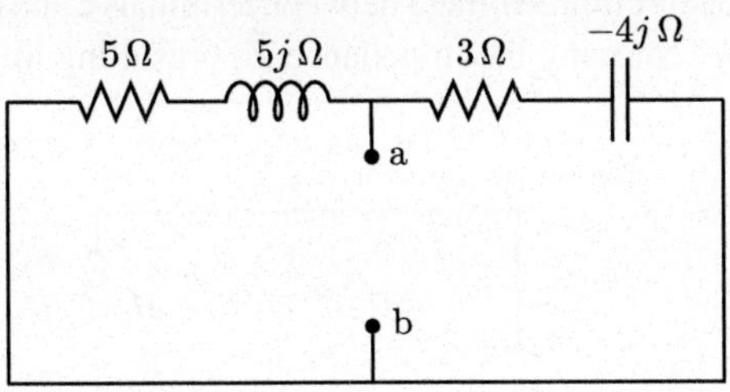

Fig. 2.141 Passive circuit

According to Fig. 2.141, the Thévenin impedance is calculated as follows:

$$\overline{Z}_{th} = (5+5j)//(3-4j) = \frac{(5+5j)\cdot(3-4j)}{5+5j+3-4j} \approx 4{,}385\angle -15{,}25^\circ\ \Omega \approx \underbrace{4{,}23}_{R_{th}} - 1{,}5j\ \Omega$$

Once the equivalent impedance from terminals a and b is calculated, then the impedance $\overline{Z}_L$ that absorbs the maximum power is as follows:

$$\overline{Z}_L = \overline{Z}^*_{eq} = 4{,}23 + 1{,}5j\ \Omega$$

On the other hand, the power absorbed by this impedance will be as follows:

$$P = \frac{U^2_{th}}{4\cdot R_{th}} = \frac{9{,}81^2}{4\cdot 4{,}23} \approx 5{,}688 \text{ W}.$$

2.40 Maximum Power Transfer

In the circuit of Fig. 2.142, it is known that when $\overline{Z}_L = 4 - 2j\,\Omega$ it receives the maximum power. Determine the Thévenin equivalent seen from the ends of $\overline{Z}_L$ and the value of R.

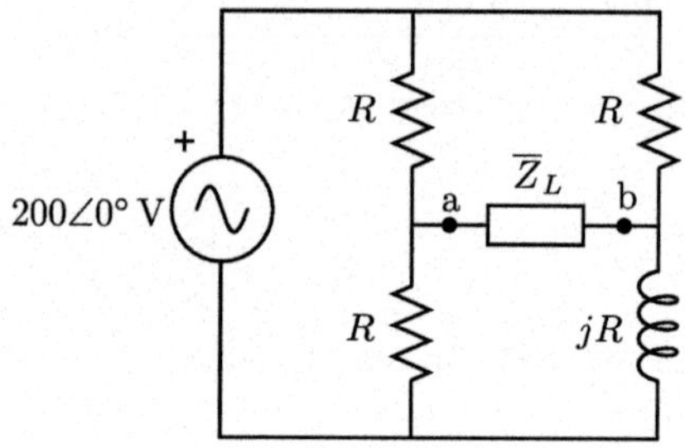

Fig. 2.142 Circuit considered in the problem under study

Solution First, the open-circuit voltage between terminals a and b, Thévenin voltage, will be calculated by removing the impedance $\overline{Z}_L$, resulting in the circuit shown in Fig. 2.143.

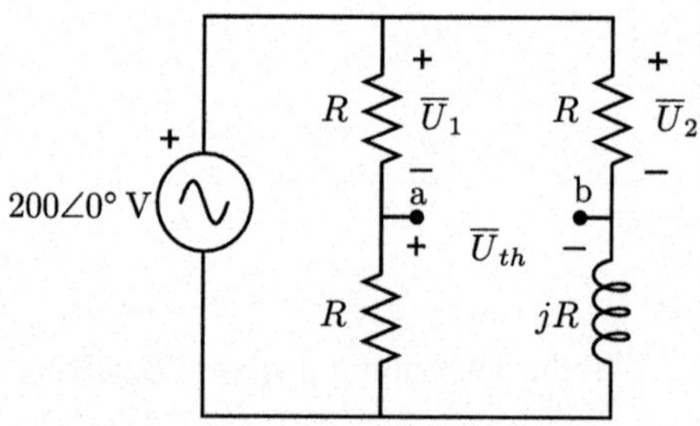

Fig. 2.143 Circuit for the calculation of Thévenin voltage

According to Fig. 2.143, applying the voltage divider concept yields the voltages $\overline{U}_1$ and $\overline{U}_2$:

$$\overline{U}_1 = 200\angle 0^\circ \cdot \frac{R}{R+R} = 100\angle 0^\circ \text{ V}$$

$$\overline{U}_2 = 200\angle 0^\circ \cdot \frac{R}{R+Rj} = 200\angle 0^\circ \cdot \frac{1}{\sqrt{2}\angle 45^\circ} = 100\sqrt{2}\angle -45^\circ \text{ V}$$

Once $\overline{U}_1$ and $\overline{U}_2$ are calculated, the Thévenin voltage, $\overline{U}_{th}$, can be calculated by applying Kirchhoff's voltage law:

$$\overline{U}_1 + \overline{U}_{th} - \overline{U}_2 = 0 \Rightarrow \overline{U}_{th} = \overline{U}_2 - \overline{U}_1 = 100\sqrt{2}\angle -45^\circ - 100\angle 0^\circ = 100\angle -90^\circ \text{ V}$$

Next, the Thévenin impedance will be calculated. For this, the circuit in Fig. 2.142 is made passive, as shown in Fig. 2.144, and the equivalent impedance between terminals a and b is calculated.

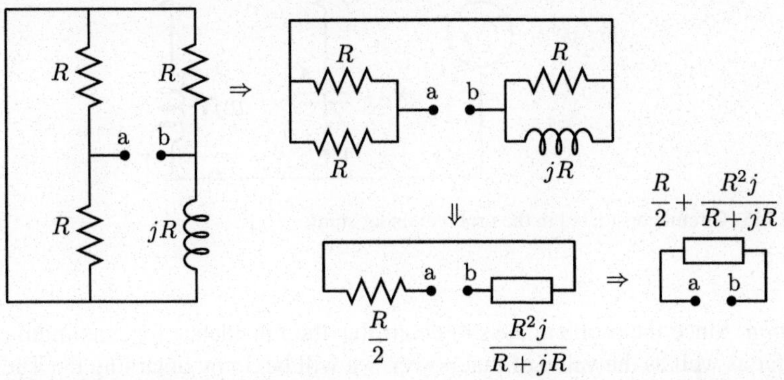

Fig. 2.144 Passive circuit

According to Fig. 2.144:

$$\overline{Z}_{th} = \frac{R}{2} + \frac{R^2 j}{R + jR} = R + 0{,}5Rj$$

On the other hand, according to the statement, it is known that the impedance $\overline{Z}_L$, once connected between terminals a and b, absorbs the maximum power. Therefore:

$$\overline{Z}_L = 4 - 2j = \overline{Z}_{th}^* \Rightarrow \overline{Z}_{th} = 4 + 2j\ \Omega$$

Equating terms yields:

$$R + 0{,}5Rj = 4 + 2j \Rightarrow R = 4\,\Omega.$$

2.41 Maximum Power Transfer

In the circuit of Fig. 2.145, the load connected between a and b consists of a variable resistance R_L and a variable capacitive reactance X_C. Determine the values of R_L and X_C that result in maximum transfer of active power to the load, as well as the value of this power, in the following cases:

1. R_L and X_C can take any value.

2. R_L can take any value, but X_C can only vary between $-8\,\Omega$ and $-2\,\Omega$.

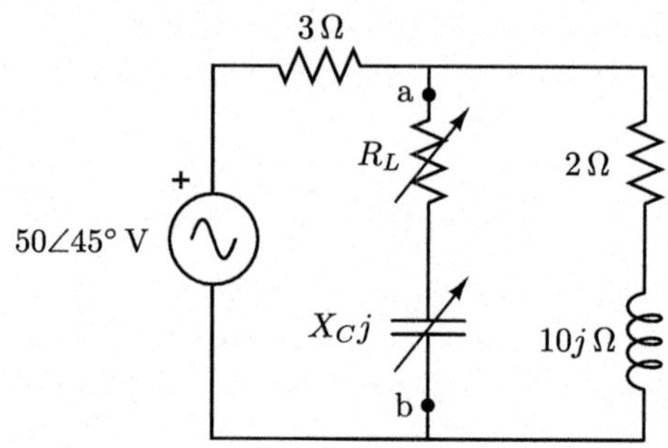

Fig. 2.145 Circuit considered in the problem under study

Solution Since the problem asks to determine the impedance for maximum power transfer as well as the value of that power, we will begin by obtaining the Thévenin equivalent from terminals a and b. First, we will calculate the open-circuit voltage between terminals a and b, Thévenin voltage, for which we will solve the circuit shown in Fig. 2.146. It can be observed that both the variable resistance R_L and the variable reactance X_C have not been considered as they are not part of the equivalent.

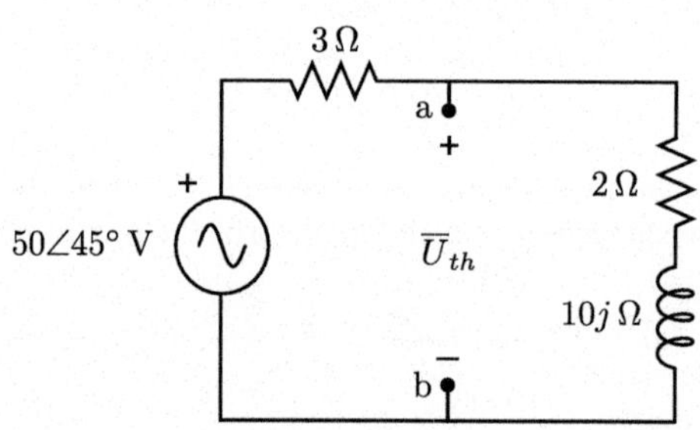

Fig. 2.146 Circuit for the calculation of Thévenin voltage

According to Fig. 2.146, applying the voltage divider concept results in:

$$\overline{U}_{th} = 50\angle 45^\circ \cdot \frac{2 + 10j}{3 + 2 + 10j} \approx 45{,}61\angle 60{,}26^\circ \text{ V}$$

Next, we calculate the equivalent impedance, Thévenin impedance, from terminals a and b. To do this, we make the circuit in Fig. 2.145 passive, as shown in Fig. 2.147.

According to Fig. 2.147, the Thévenin impedance corresponds to the parallel connection of the $3\,\Omega$ resistance and the $2 + 10j\,\Omega$ impedance:

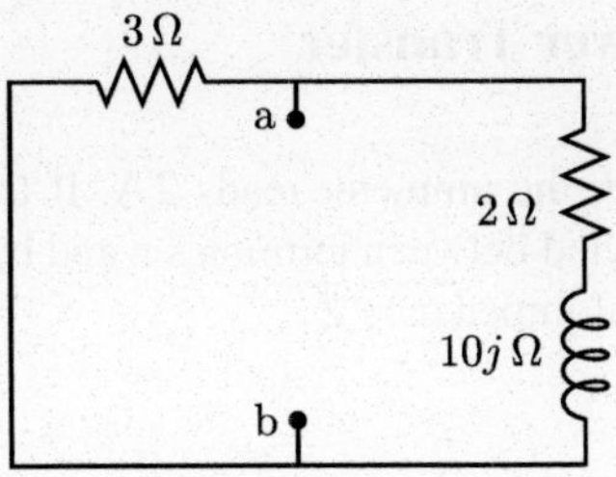

Fig. 2.147 Passive circuit

$$\overline{Z}_{th} = \frac{3 \cdot (2 + 10j)}{3 + 2 + 10j} \approx \underbrace{2{,}64}_{R_{th}} + \underbrace{0{,}72j}_{X_{th}j} \ \Omega$$

1. Once the equivalent impedance from terminals a and b is calculated, the values of R_L and X_C that result in maximum power absorption are as follows:

$$R_L + X_C j = \overline{Z}_{th}^* = 2{,}64 - 0{,}72j \ \Omega$$

On the other hand, the power absorbed by this impedance is:

$$P = \frac{U_{th}^2}{4 \cdot R_{th}} = \frac{45{,}61^2}{4 \cdot 2{,}64} \approx 197 \text{ W}$$

2. In this case, R_L can take any value, but X_C is limited to a range of values. Thus, R_L is as follows:

$$R_L^2 = R_{th}^2 + (X_{th} + X_C)^2 = 2{,}64^2 + (0{,}72 + X_C)^2$$
$$\Rightarrow \ R_L = \sqrt{2{,}64^2 + (0{,}72 + X_C)^2}$$

where $X_C \in [-8, -2]$.
In this case, the value of the power absorbed by this impedance is as follows:

$$P = \frac{U_{th}^2}{2 \cdot (R_{th} + R_L)} = \frac{45{,}61^2}{2 \cdot (2{,}64 + \sqrt{2{,}64^2 + (0{,}72 + X_C)^2})}$$

It can be verified that the value of X_C within the interval $[-8, -2]$ that maximizes the power is $X_C = -2\,\Omega$. Consequently:

$$P = \frac{45{,}61^2}{2 \cdot (2{,}64 + \sqrt{2{,}64^2 + (0{,}72 + X_C)^2})}$$
$$= \frac{45{,}61^2}{2 \cdot (2{,}64 + \sqrt{2{,}64^2 + (0{,}72 - 2)^2})} \approx 186{,}61 \text{ W}.$$

2.42 Maximum Power Transfer

In the circuit of Fig. 2.148, the ammeter reads 2 A. If the impedance that absorbs maximum power is connected between terminals a and b, and it is known to receive 32 W, calculate the value of impedance $\overline{Z}$.

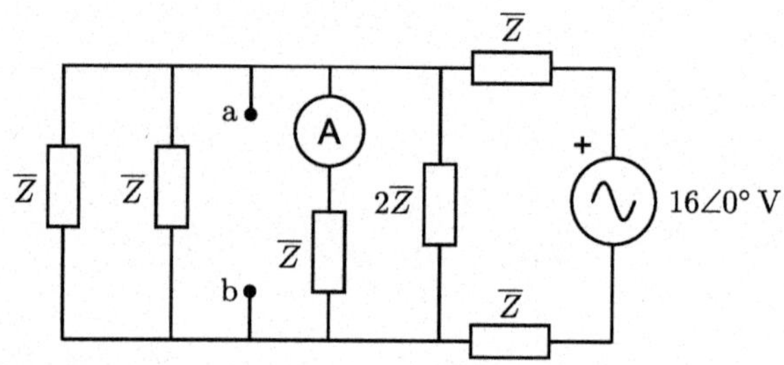

Fig. 2.148 Circuit considered in the problem under study

Solution First, the Thévenin voltage between terminals a and b will be obtained, which corresponds, according to the circuit in Fig. 2.149, to calculating the voltage between terminals a and b. This circuit can be simplified by associating impedances in parallel, as shown in Fig. 2.150.

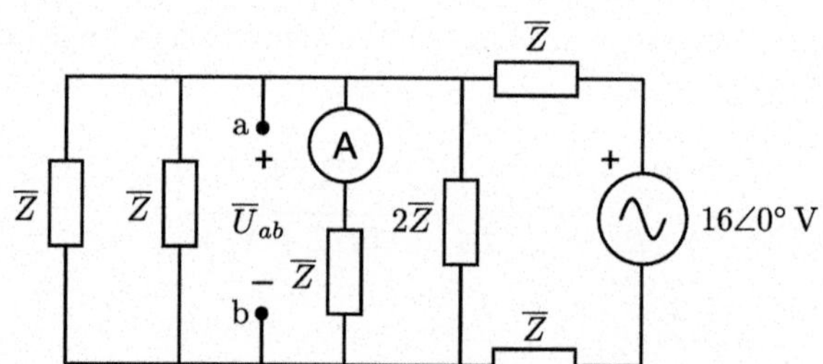

Fig. 2.149 Circuit considered in the problem under study

According to Fig. 2.150, applying the voltage divider concept yields:

$$\overline{U}_{th} = \overline{U}_{ab} = 16\angle 0^\circ \frac{\dfrac{2\overline{Z}}{7}}{\dfrac{2\overline{Z}}{7} + 2\overline{Z}} = 2\angle 0^\circ \text{ V}$$

On the other hand, according to the statement, the reading of the ammeter is 2 A; therefore, according to Fig. 2.149:

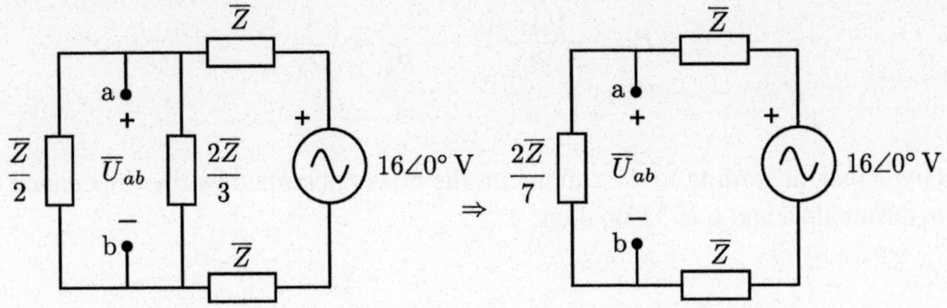

Fig. 2.150 Simplified circuit

$$|\overline{U}_{ab}| = U_{ab} = 2 \cdot Z \;\Rightarrow\; Z = \frac{U_{ab}}{2} = \frac{2}{2} = 1\,\Omega$$

Next, the Thévenin impedance between terminals a and b will be obtained, for which the circuit in Fig. 2.148 will be made passive, resulting in the circuit of Fig. 2.151.

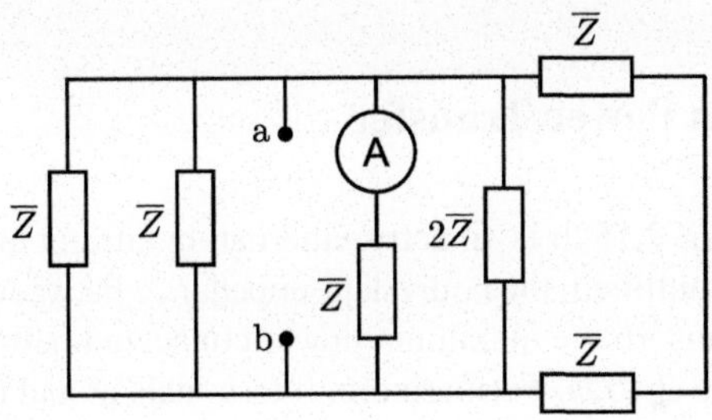

Fig. 2.151 Passive circuit

According to Fig. 2.151, the Thévenin impedance is calculated as follows:

$$\overline{Z}_{th} = \frac{1}{\dfrac{1}{\overline{Z}} + \dfrac{1}{\overline{Z}} + \dfrac{1}{\overline{Z}} + \dfrac{1}{2\overline{Z}} + \dfrac{1}{2\overline{Z}}} = \frac{\overline{Z}}{4} = \underbrace{\frac{R_Z}{4}}_{R_{th}} + j\frac{X_Z}{4}$$

The impedance $\overline{Z}_L$ that, once connected between a and b, would absorb the maximum power is

$$\overline{Z}_L = \overline{Z}_{th}^{*} = \frac{R_Z}{4} - j\frac{X_Z}{4}$$

and the power absorbed is:

$$P = \frac{U_{th}^2}{4 \cdot R_{th}} = \frac{2^2}{4 \cdot \frac{R_Z}{4}} = \frac{4}{R_Z}$$

Given that, according to the statement, the power absorbed by the impedance $\overline{Z}_L$ from terminals a and b is 32 W, then:

$$P = \frac{4}{R_Z} = 32 \Rightarrow R_Z = \frac{4}{32} = \frac{1}{8}\,\Omega$$

Finally, the value of X_Z is calculated as:

$$X_Z = \pm\sqrt{Z^2 - R_Z^2} = \pm\sqrt{1^2 - \left(\frac{1}{8}\right)^2} \approx \pm 0{,}992\,\Omega$$

Therefore:

$$\overline{Z} = \frac{1}{8} \pm 0{,}992j \approx 1\angle \pm 82{,}82°\,\Omega.$$

2.43 Maximum Power Transfer

The AC circuit of Fig. 2.152b is an active alternating current circuit. If all its independent sources are nullified, the equivalent impedance between a' and b' would be the one that would absorb the maximum power connected between terminals a and b of the circuit in Fig. 2.152a. Furthermore, if terminals a' and b' of Fig. 2.152b are short-circuited, then $\overline{I}_{a'b'} = 10\angle 0°$ A. Calculate the voltage between terminals a and b if both circuits were connected.

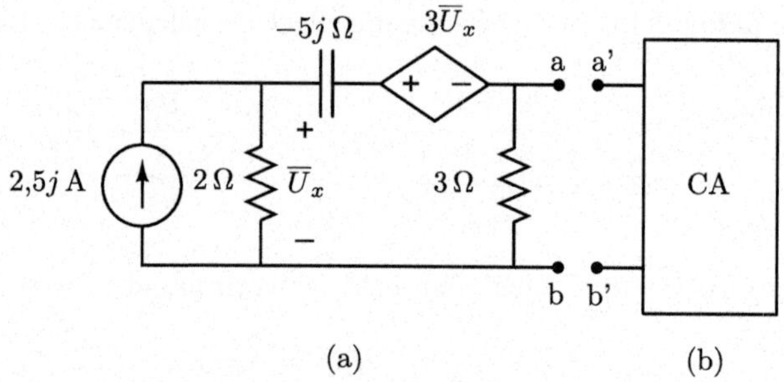

Fig. 2.152 Circuit considered in the problem under study

Solution First, the open-circuit voltage between terminals a and b of the circuit in Fig. 2.152a, Thévenin's voltage, will be calculated. To do this, the circuit shown in Fig. 2.153 will be solved.

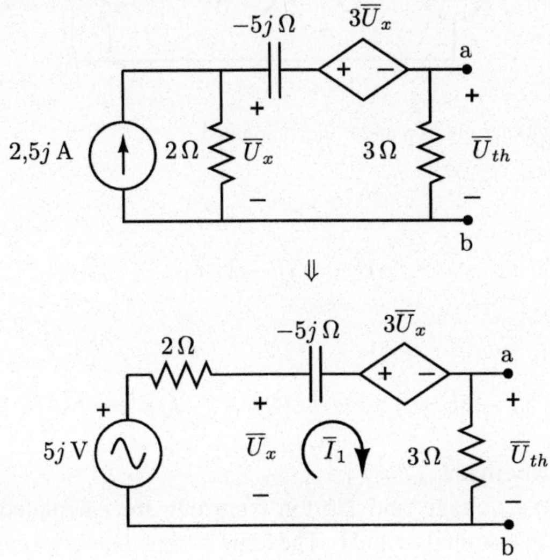

Fig. 2.153 Circuit to be solved

According to Fig. 2.153, using the mesh method yields:

$$(2 - 5j + 3) \cdot \overline{I}_1 = 5j - 3\overline{U}_x$$

On the other hand:

$$\overline{U}_x = 5j - 2 \cdot \overline{I}_1$$

Consequently:

$$(2 - 5j + 3) \cdot \overline{I}_1 = 5j - 3\overline{U}_x = 5j - 3 \cdot (5j - 2 \cdot \overline{I}_1) \;\Rightarrow\; \overline{I}_1 \approx 1{,}96\angle 11{,}31^\circ\,\text{A}$$

Once $\overline{I}_1$ is obtained, Thévenin's voltage is as follows:

$$\overline{U}_{th} = 3 \cdot \overline{I}_1 = 3 \cdot 1{,}96\angle 11{,}31^\circ = 5{,}88\angle 11{,}31^\circ\,\text{V}$$

Next, Norton's current is calculated, which is obtained by short-circuiting terminals a and b, as shown in Fig. 2.154.

According to Fig. 2.154, applying the mesh method yields (note that the $3\,\Omega$ resistance is short-circuited and therefore does not participate in the resolution):

$$(2 - 5j) \cdot \overline{I}_1 = 5j - 3\overline{U}_x$$

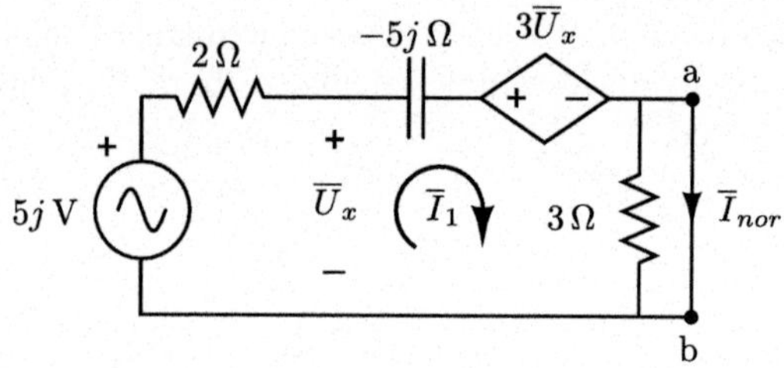

Fig. 2.154 Circuit with terminals a and b short-circuited

On the other hand:

$$\overline{U}_x = 5j - 2 \cdot \overline{I}_1$$

Consequently:

$$(2 - 5j) \cdot \overline{I}_1 = 5j - 3\overline{U}_x = 5j - 3 \cdot (5j - 2 \cdot \overline{I}_1) \Rightarrow \overline{I}_1 \approx 1{,}56\angle 38{,}66^\circ \text{ A}$$

It is easy to verify that $\overline{I}_{nor} = \overline{I}_1$.

Once Thévenin's voltage and Norton's current are calculated, the equivalent impedance between terminals a and b, Thévenin's impedance, is obtained as follows:

$$\overline{Z}_{th} = \frac{\overline{U}_{th}}{\overline{I}_{nor}} = \frac{5{,}88\angle 11{,}31^\circ}{1{,}56\angle 38{,}66^\circ} \approx 3{,}76\angle - 27{,}35^\circ \ \Omega$$

From circuit (b) of Fig. 2.152, it is known that its equivalent impedance is the one that would absorb the maximum power once connected between terminals a and b of circuit (a). That is, the equivalent impedance of circuit (b) bears the following relation with the equivalent impedance of circuit (a):

$$\overline{Z}_{th} \mid_{a'b'} = \overline{Z}^*_{th} = 3{,}76\angle 27{,}35^\circ \ \Omega$$

Furthermore, according to the statement, if terminals a' and b' of circuit (b) are short-circuited, then the current flowing through the short circuit is $10\angle 0^\circ$ A. This current is the Norton current of circuit (b).

Once the equivalents of both circuits are calculated, the circuit in Fig. 2.155 is obtained. It can be observed that Norton's equivalent has been used for the representation of each circuit.

According to Fig. 2.153, using the node method yields:

$$\left(\frac{1}{3{,}76\angle - 27{,}35^\circ} + \frac{1}{3{,}76\angle 27{,}35^\circ}\right) \cdot \overline{U}_{ab} = 1{,}56\angle 38{,}66^\circ + 10\angle 0^\circ$$

Solving yields the voltage $\overline{U}_{ab}$:

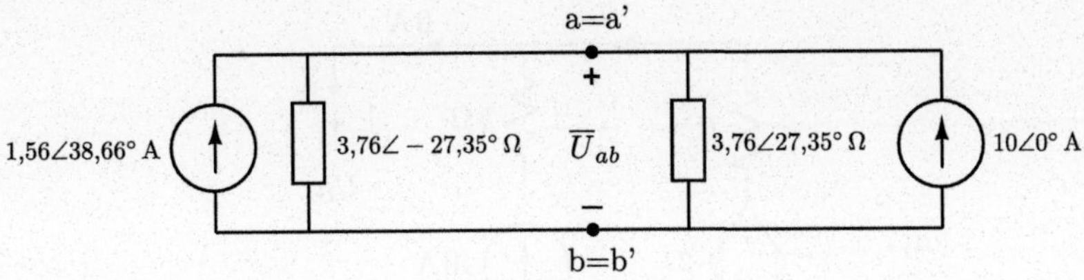

Fig. 2.155 Equivalent circuit

$$\overline{U}_{ab} \approx 23{,}83\angle 4{,}96^\circ \text{ V}.$$

2.44 Maximum Power Transfer

Determine the value of R for the 3 Ω resistance to absorb the maximum power from the rest of the circuit in Fig. 2.156. Also, determine the value of said power.

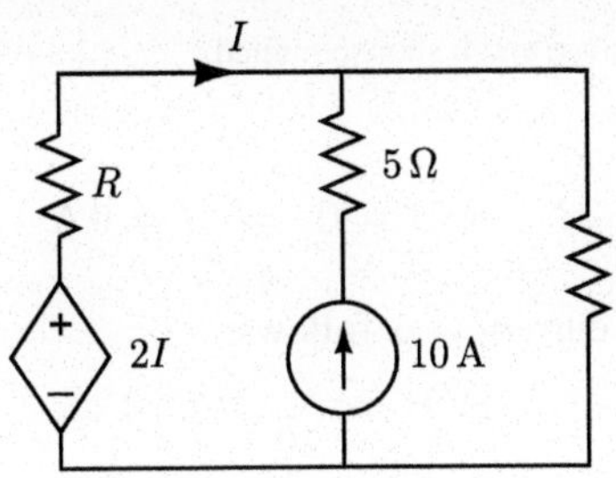

Fig. 2.156 Circuit considered in the problem under study

Solution First, the Thévenin equivalent will be obtained from terminals a and b of the circuit in Fig. 2.156. The circuit for the calculation of Thévenin's voltage is shown in Fig. 2.157. It can be seen that the 3 Ω resistance has not been considered since it is not part of the equivalent.

According to Fig. 2.157, it is easy to verify that $I = -10$ A. Therefore:

$$U_{th} = 2I - R \cdot I = 10 \cdot R - 20$$

Next, Norton's current will be calculated, for which the circuit in Fig. 2.158 will be taken into account. According to Fig. 2.158, applying Kirchhoff's current law yields:

$$I + 10 = I_{nor}$$

On the other hand:

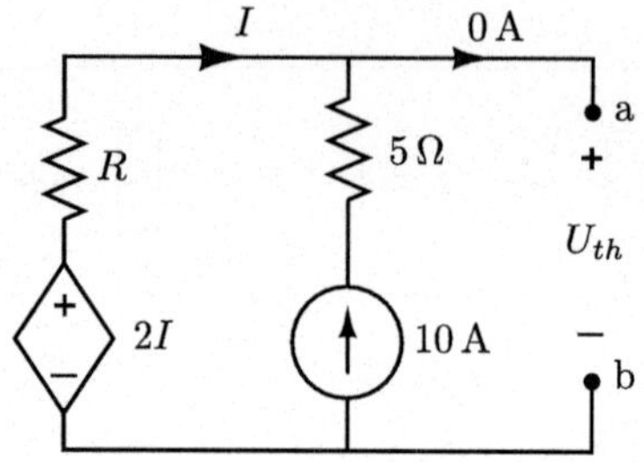

Fig. 2.157 Circuit to be solved

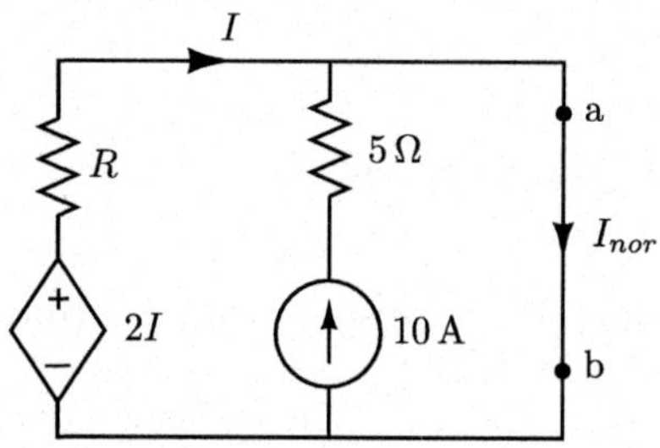

Fig. 2.158 Circuit with terminals a and b short-circuited

$$-2I + R \cdot I = 0 \;\Rightarrow\; I = 0\,\text{A}$$

Consequently, Norton's current is as follows:

$$I_{nor} = 10\,\text{A}$$

Once Thévenin's voltage and Norton's current are calculated, the equivalent resistance between terminals a and b is obtained as follows:

$$R_{eq} = \frac{U_{th}}{I_{nor}} = \frac{10 \cdot R - 20}{10} = R - 2$$

For the 3 Ω resistance to absorb the maximum power from the circuit, $R_{eq} = 3$ is necessary, therefore:

$$R_{eq} = R - 2 = 3 \;\Rightarrow\; R = 5\,\Omega$$

On the other hand, the power absorbed by the 3 Ω resistance from the circuit is calculated as follows:

$$P = \frac{U_{th}^2}{4 \cdot 3} = \frac{(10 \cdot R - 20)^2}{4 \cdot 3} = \frac{(10 \cdot 5 - 20)^2}{4 \cdot 3} = 75\,\text{W}.$$

2.45 Maximum Power Transfer

In the circuit of Fig. 2.159, it is known that $\overline{Z}$ is the impedance that absorbs the maximum possible power from circuit A. Determine the reading of the ammeter and voltmeter.

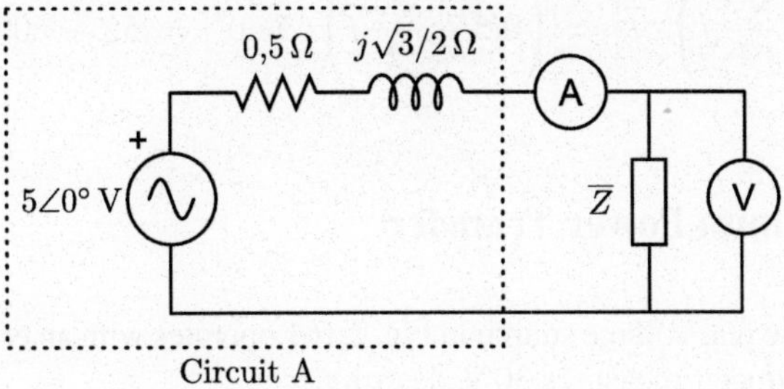

Fig. 2.159 Circuit considered in the problem under study

Solution According to Fig. 2.159, it is easy to verify that the equivalent impedance of circuit A, that is, from the terminals of impedance $\overline{Z}$, is as follows:

$$\overline{Z}_{eq} = 0{,}5 + \frac{\sqrt{3}}{2} j \; \Omega$$

According to the statement, impedance $\overline{Z}$ is the one that absorbs the maximum power from circuit A, therefore:

$$\overline{Z} = \overline{Z}^{*}_{eq} = 0{,}5 - \frac{\sqrt{3}}{2} j \; \Omega$$

In this way, the resulting circuit is shown in Fig. 2.160.

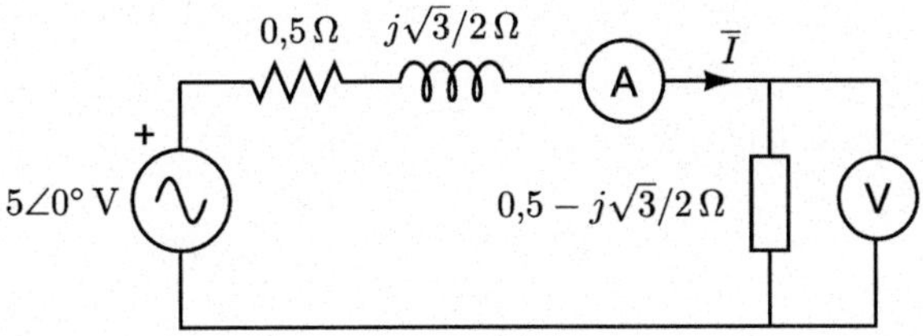

Fig. 2.160 Resulting circuit

According to Fig. 2.160:

$$\overline{I} = \frac{5\angle 0^\circ}{0{,}5 + \frac{\sqrt{3}}{2}j + 0{,}5 - \frac{\sqrt{3}}{2}j} = 5\angle 0^\circ \text{ A}$$

Therefore, the reading of the ammeter is 5 A.
The reading of the voltmeter is calculated as follows:

$$\left|\left(0{,}5 - \frac{\sqrt{3}}{2}j\right) \cdot \overline{I}\right| = \left|\left(0{,}5 - \frac{\sqrt{3}}{2}j\right) \cdot 5\angle 0^\circ\right| = |5\angle - 60^\circ| = 5 \text{ V}.$$

2.46 Maximum Power Transfer

Knowing that the real voltage source in Fig. 2.161 operates with an efficiency of 50% and that the voltmeter measures 30 V, determine:

1. The voltage U_g of the source.
2. The equivalent resistance of the resistive circuit.

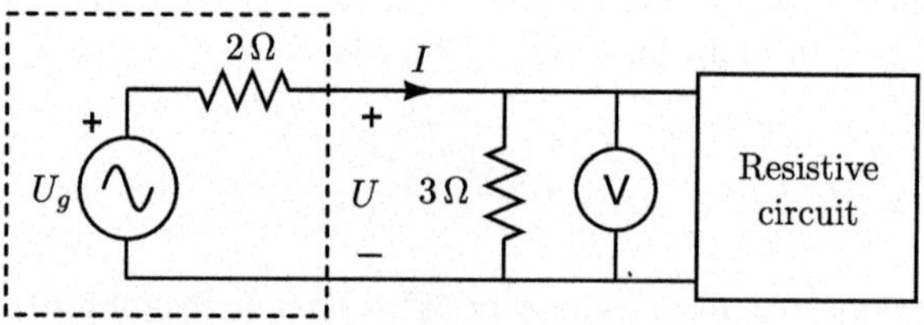

Fig. 2.161 Circuit considered in the problem under study

Solution 1. The efficiency of the real voltage source is as follows:

$$\eta = \frac{U \cdot I}{U_g \cdot I} \cdot 100 = \frac{U}{U_g} \cdot 100$$

where U is the voltage measured by the voltmeter.
Therefore:

$$U_g = \frac{U}{\eta} \cdot 100 = \frac{30}{50} \cdot 100 = 60 \text{ V}$$

2. Since the efficiency of the real voltage source is 50%, it implies that the resistance of said real source is equal to the equivalent resistance connected to said source. Consequently:

$$R_{eq} = 2 \;\Rightarrow\; 3//R_{CR} = 2 \;\Rightarrow\; \frac{3 \cdot R_{CR}}{3 + R_{CR}} = 2$$

Solving for and solving yields the value of the resistance of the resistive circuit:

$$R_{CR} = 6\,\Omega.$$

2.47 Maximum Power Transfer

In the circuit of Fig. 2.162, determine the value of G_L so that it absorbs maximum power. Also, determine the value of that power.

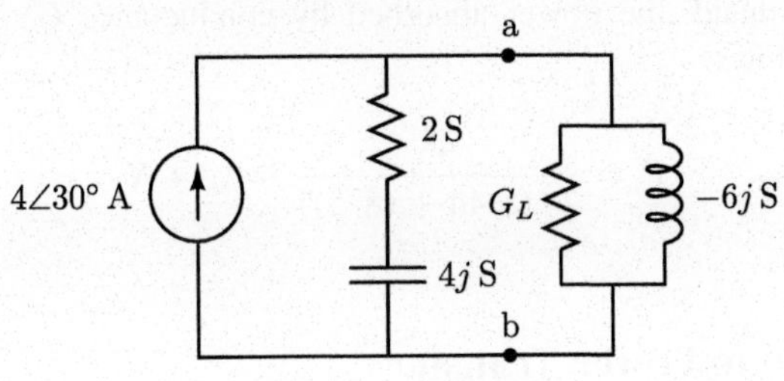

Fig. 2.162 Circuit considered in the problem under study

Solution First, we will calculate the equivalent impedance between terminals a and b of the circuit in Fig. 2.162, without considering the branch formed by the conductance G_L and the inductor:

$$\overline{Z}_{eq} = \frac{1}{2} + \frac{1}{4j} = 0{,}5 - 0{,}25j\;\Omega$$

Once the equivalent impedance is obtained, we are interested in expressing it in the form of admittance as follows:

$$\overline{Y}_{eq} = \frac{1}{\overline{Z}_{eq}} = \frac{1}{0{,}5 - 0{,}25j} = \underbrace{\frac{8}{5}}_{G_{eq}} + \underbrace{\frac{4}{5}}_{B_{eq}}\,j\,\text{S}$$

Thus, the circuit of Fig. 2.162 is as shown in Fig. 2.163.

According to Fig. 2.163 and considering the maximum power transfer theorem for variable G_L and fixed B_L, the value of conductance G_L that absorbs maximum power is obtained as follows:

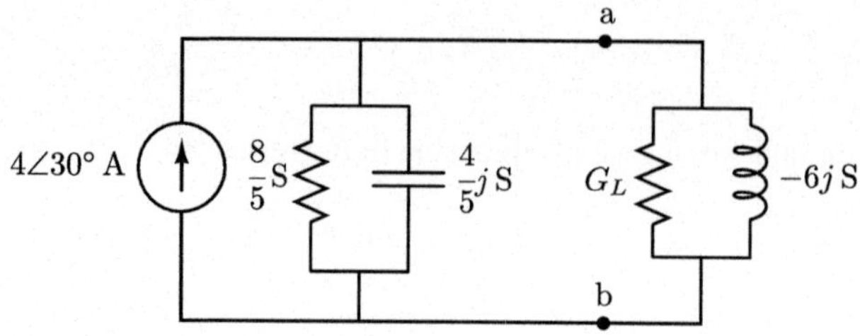

Fig. 2.163 Equivalent circuit

$$G_L^2 = G_{eq}^2 + (B_{eq} + B_L)^2 = \left(\frac{8}{5}\right)^2 + \left(\frac{4}{5} - 6\right)^2 = 29{,}6\,\mathrm{S}^2 \Rightarrow G_L \approx 5{,}44\,\mathrm{S}$$

On the other hand, the power absorbed by conductance G_L in the circuit is calculated as follows:

$$P = \frac{4^2}{2 \cdot (5{,}44 + (8/5))} \approx 1{,}14\,\mathrm{W}.$$

2.48 Maximum Power Transfer

In the circuit of Fig. 2.164, determine the impedance $\overline{Z}$ so that it absorbs maximum power from the circuit. Also, determine the value of that power.

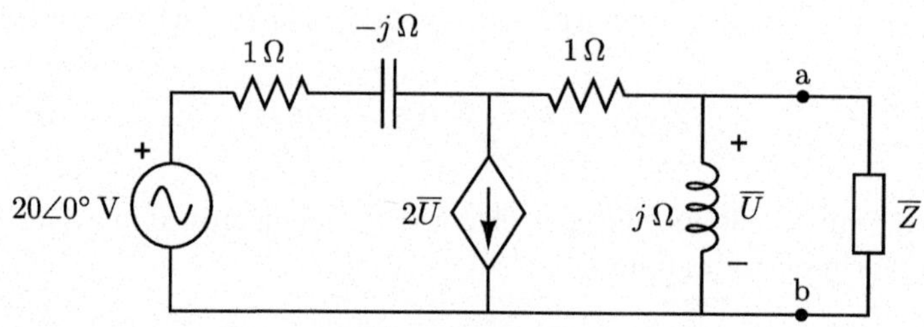

Fig. 2.164 Circuit considered in the problem under study

Solution To calculate the impedance $\overline{Z}$ that absorbs maximum power and the value of that power, it is necessary to obtain the Thévenin equivalent from terminals a and b.

First, we will calculate the open-circuit voltage between terminals a and b, Thévenin voltage. To do this, we will solve the circuit shown in Fig. 2.165. It can be seen that the impedance $\overline{Z}$ has not been considered because it is not part of the equivalent.

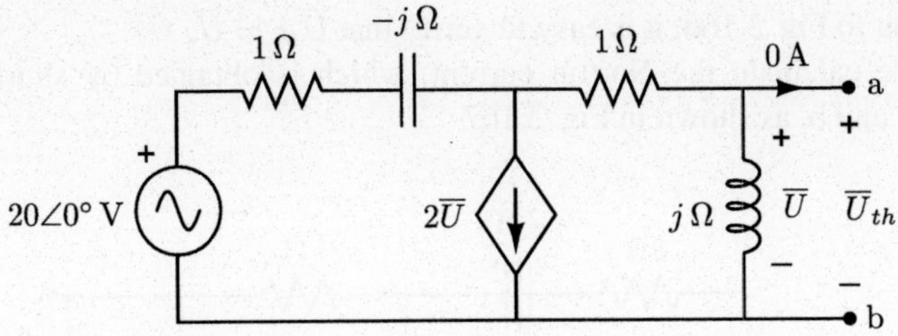

Fig. 2.165 Circuit to be solved for Thévenin voltage

The circuit in Fig. 2.165 can be simplified by replacing the real voltage source with its equivalent real current source as shown in Fig. 2.166. According to Fig. 2.166, applying the nodal analysis method:

$$\left(\frac{1}{1-j}+\frac{1}{1+j}\right)\cdot\overline{U}_1=\frac{20\angle 0^\circ}{1-j}-2\overline{U}$$

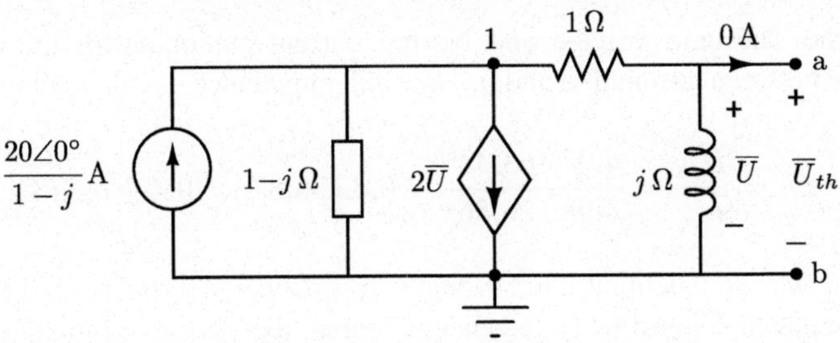

Fig. 2.166 Simplified circuit

On the other hand, applying the voltage divider concept:

$$\overline{U}=\overline{U}_1\cdot\frac{j}{1+j}\quad\Rightarrow\quad\overline{U}_1=\overline{U}\cdot\frac{1+j}{j}$$

Consequently:

$$\left(\frac{1}{1-j}+\frac{1}{1+j}\right)\cdot\overline{U}\cdot\frac{1+j}{j}=\frac{20\angle 0^\circ}{1-j}-2\overline{U}$$

Solving for and solving, we obtain the value of $\overline{U}$:

$$\overline{U}\approx 4{,}47\angle 63{,}44^\circ\ \text{V}$$

Referring to Fig. 2.166, it is easy to verify that $\overline{U}_{th} = \overline{U}$.

Next, we calculate the Norton current, which is obtained by short-circuiting terminals a and b, as shown in Fig. 2.167.

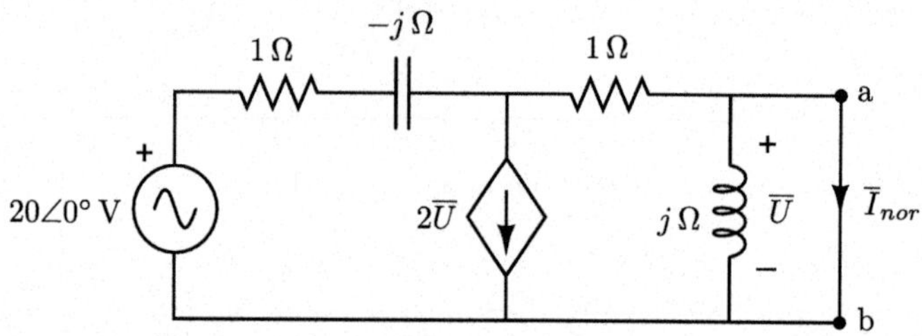

Fig. 2.167 Circuit with terminals a and b short-circuited

According to Fig. 2.167, we can see that $\overline{U} = 0$, therefore:

$$\overline{I}_{nor} = \frac{20\angle 0^\circ}{1 - j + 1} \approx 8{,}94\angle 26{,}56^\circ \text{ A}$$

Once the Thévenin voltage and Norton current are obtained, the equivalent impedance between terminals a and b, Thévenin impedance, is calculated as follows:

$$\overline{Z}_{th} = \frac{\overline{U}_{th}}{\overline{I}_{nor}} = \frac{4{,}47\angle 63{,}44^\circ}{8{,}94\angle 26{,}56^\circ} \approx 0{,}5\angle 36{,}87^\circ = 0{,}4 + 0{,}3j\ \Omega$$

Another way to calculate the Thévenin impedance is from the passive circuit. In this case, as the circuit has a dependent source, the Thévenin impedance cannot be obtained by impedance association. Instead, a test source connected between terminals a and b to the passive circuit could be used, as shown in Fig. 2.168.

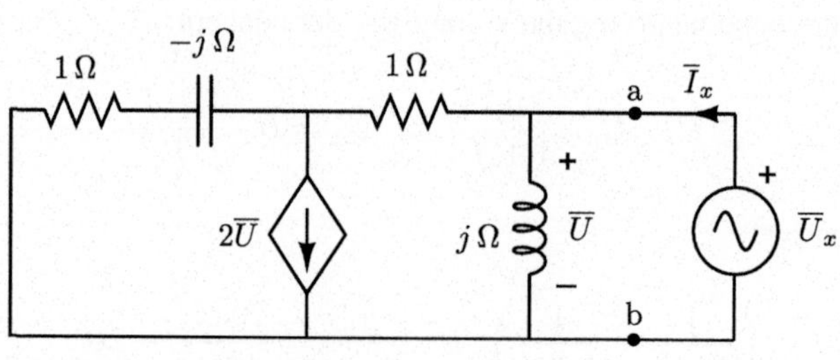

Fig. 2.168 Circuit with a test voltage source

The circuit in Fig. 2.168 can be simplified by replacing the real current source with its equivalent voltage source as shown in Fig. 2.169. Additionally, the currents $\overline{I}_1$ and $\overline{I}_2$ are indicated to facilitate the resolution.

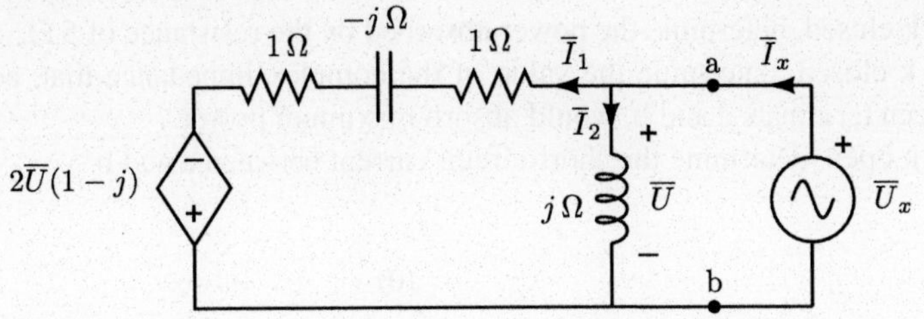

Fig. 2.169 Simplified circuit

According to Fig. 2.169, it can be observed that $\overline{U} = \overline{U}_x$. On the other hand:

$$\overline{I}_1 = \frac{\overline{U}_x + 2\overline{U}_x(1-j)}{2-j}$$

$$\overline{I}_2 = \frac{\overline{U}_x}{j}$$

The current $\overline{I}_x$ is obtained by applying Kirchhoff's current law:

$$\overline{I}_x = \overline{I}_1 + \overline{I}_2 = \frac{\overline{U}_x + 2\overline{U}_x(1-j)}{2-j} + \frac{\overline{U}_x}{j} = \frac{4+2j}{(2-j)\cdot j}\cdot \overline{U}_x$$

Solving for and solving yields:

$$\overline{Z}_{th} = \frac{\overline{U}_x}{\overline{I}_x} = \frac{(2-j)\cdot j}{4+2j} = 0{,}4 + 0{,}3j\ \Omega$$

Once the Thévenin impedance is calculated from terminals a and b, the impedance $\overline{Z}$ that absorbs maximum power is as follows:

$$\overline{Z} = \overline{Z}_{th}^{*} = \underbrace{0{,}4}_{R_{th}}\ -0{,}3j\ \Omega$$

On the other hand, the power absorbed by this impedance will be:

$$P_{max} = \frac{U_{th}^2}{4\cdot R_{th}} = \frac{4,47^2}{4\cdot 0{,}4} \approx 12{,}48\ \mathrm{W}.$$

2.49 Maximum Power Transfer

In the circuit of Fig. 2.170 the following is requested:

1. With k closed, determine the power absorbed by the resistance of 5 Ω.
2. With k closed, determine the value of the complex impedance that, connected between terminals a and b, would absorb maximum power.
3. With k open, determine the short-circuit current between a and b.

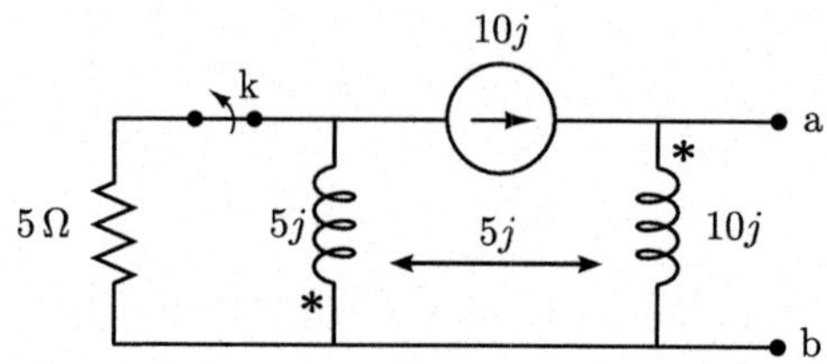

Fig. 2.170 Circuit considered in the problem under study

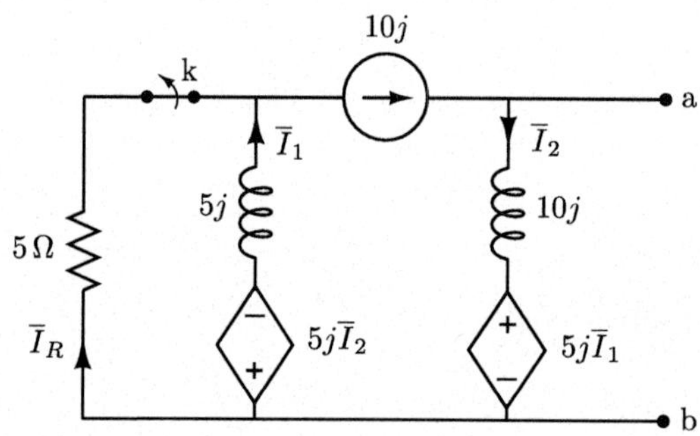

Fig. 2.171 Equivalent circuit for coupled coils

Solution First, the coupled coils will be replaced by their equivalent model with dependent voltage sources in the phasor domain as shown in Fig. 2.171.

1. According to Fig. 2.171, it is easy to verify that $\overline{I}_2 = 10j$ A. On the other hand, applying Kirchhoff's current law results in:

$$\overline{I}_R + \overline{I}_1 = 10j \Rightarrow \overline{I}_1 = 10j - \overline{I}_R$$

Finally, applying Kirchhoff's voltage law:

$$5 \cdot \overline{I}_R - 5j \cdot \overline{I}_1 - 5j\overline{I}_2 = 0$$

Substituting values and solving yields:

$$5 \cdot \overline{I}_R - 5j \cdot (10j - \overline{I}_R) - 5j \cdot 10j = 0 \Rightarrow \overline{I}_R = 10\sqrt{2}\angle 135^\circ \text{ A}$$

Once the current flowing through the resistance is obtained, the power absorbed by it is as follows:

$$P = 5 \cdot (10\sqrt{2})^2 = 1\,000\,\text{W}$$

2. The calculation of the complex impedance that, connected between terminals a and b, absorbs maximum power involves knowing the equivalent impedance of the circuit between those terminals. The calculation of the equivalent impedance can be carried out in several alternative ways. In this case, for didactic reasons, all of them will be shown. First, the Thévenin voltage, i.e., the voltage that results when terminals a and b are left open, will be obtained. To do this, the circuit of Fig. 2.172 will be solved.
Observing the circuit of Fig. 2.172, it can be verified that it is exactly the same circuit as that of Fig. 2.171 and, therefore, the results obtained in the previous section can be used. According to Fig. 2.172:

$$\overline{U}_{th} = 10j \cdot \overline{I}_2 + 5j \cdot \overline{I}_1$$

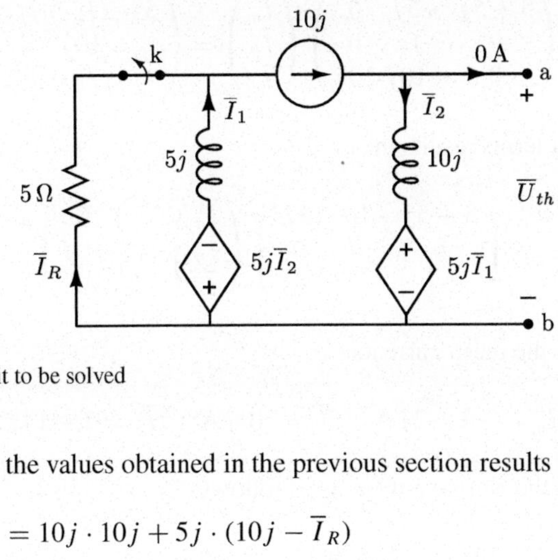

Fig. 2.172 Circuit to be solved

Substituting the values obtained in the previous section results in:

$$\begin{aligned}\overline{U}_{th} &= 10j \cdot 10j + 5j \cdot (10j - \overline{I}_R) \\ &= 10j \cdot 10j + 5j \cdot (10j - 10\sqrt{2}\angle 135^\circ) = -100 + 50j\,\text{V}\end{aligned}$$

Next, the Norton current, which is obtained by short-circuiting terminals a and b, will be calculated, as shown in Fig. 2.173.
According to Fig. 2.173, applying the mesh analysis method:

$$\begin{pmatrix} 5+5j & -5j & 0 \\ 0 & 1 & 0 \\ 0 & -10j & 10j \end{pmatrix} \begin{pmatrix} \overline{I}_a \\ \overline{I}_b \\ \overline{I}_c \end{pmatrix} = \begin{pmatrix} 5j \cdot \overline{I}_2 \\ 10j \\ 5j \cdot \overline{I}_1 \end{pmatrix}$$

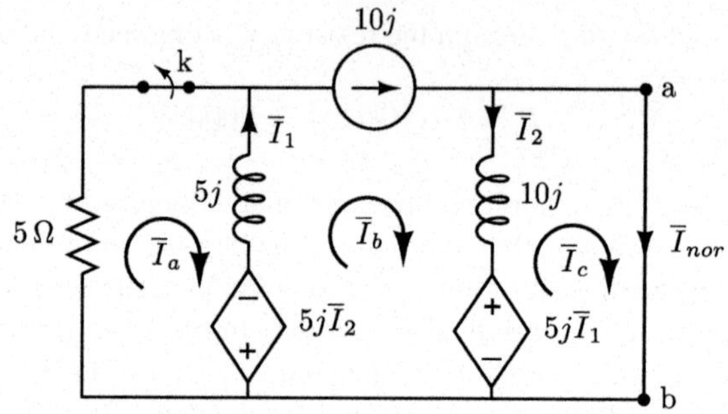

Fig. 2.173 Circuit with terminals a and b short-circuited

On the other hand:

$$\overline{I}_1 = \overline{I}_b - \overline{I}_a \ ; \ \overline{I}_2 = \overline{I}_b - \overline{I}_c$$

Substituting values

$$\begin{pmatrix} 5+5j & -5j & 0 \\ 0 & 1 & 0 \\ 0 & -10j & 10j \end{pmatrix} \begin{pmatrix} \overline{I}_a \\ \overline{I}_b \\ \overline{I}_c \end{pmatrix} = \begin{pmatrix} 5j \cdot (\overline{I}_b - \overline{I}_c) \\ 10j \\ 5j \cdot (\overline{I}_b - \overline{I}_a) \end{pmatrix}$$

and reordering terms results in:

$$\begin{pmatrix} 5+5j & -10j & 5j \\ 0 & 1 & 0 \\ 5j & -15j & 10j \end{pmatrix} \begin{pmatrix} \overline{I}_a \\ \overline{I}_b \\ \overline{I}_c \end{pmatrix} = \begin{pmatrix} 0 \\ 10j \\ 0 \end{pmatrix}$$

Solving yields the mesh currents:

$$\overline{I}_a = -4 + 2j \text{ A} \ ; \ \overline{I}_b = 10j \text{ A} \ ; \ \overline{I}_c = 2 + 14j \text{ A}$$

Consequently, the Norton current is as follows:

$$\overline{I}_{nor} = \overline{I}_c = 2 + 14j \text{ A}$$

Once the Thévenin voltage and the Norton current are calculated, the Thévenin impedance between terminals a and b is obtained as follows:

$$\overline{Z}_{th} = \frac{\overline{U}_{th}}{\overline{I}_{nor}} = \frac{-100 + 50j}{2 + 14j} = 2{,}5 + 7{,}5j \ \Omega$$

An alternative way to obtain the equivalent impedance is by using a test source connected between terminals a and b to the passive circuit as shown in Fig. 2.174.

In this case, as the circuit has a dependent source, the equivalent impedance cannot be obtained by impedance association in the passive circuit.
According to Fig. 2.174:

$$\overline{U}_x - 5j\overline{I}_1 - 10j \cdot \overline{I}_x = 0$$

On the other hand:

$$5j \cdot \overline{I}_x + 5j \cdot \overline{I}_1 + 5 \cdot \overline{I}_1 = 0$$

Substituting values and solving yields the relationship between $\overline{U}_x$ and $\overline{I}_x$:

$$\overline{Z}_{th} = \frac{\overline{U}_x}{\overline{I}_x} = 10j + \frac{25}{5+5j} = 2{,}5 + 7{,}5j\ \Omega$$

Once the Thévenin impedance is calculated from terminals a and b, the impedance $\overline{Z}$ that, connected between those terminals, absorbs maximum power is as follows:

$$\overline{Z} = \overline{Z}_{th}^{*} = 2{,}5 - 7{,}5j\ \Omega$$

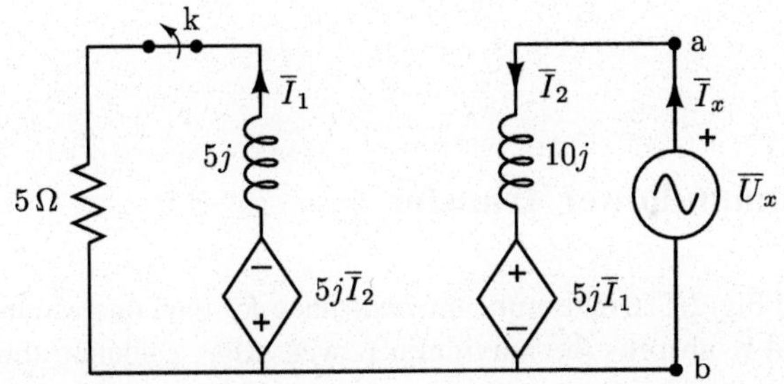

Fig. 2.174 Circuit with a test voltage source

3. With k open, the short-circuit current between terminals a and b is obtained from the circuit of Fig. 2.175. According to Fig. 2.175, applying the mesh analysis method:

$$\begin{pmatrix} 1 & 0 \\ -10j & 10j \end{pmatrix} \begin{pmatrix} \overline{I}_a \\ \overline{I}_b \end{pmatrix} = \begin{pmatrix} 10j \\ 5j \cdot \overline{I}_1 \end{pmatrix}$$

On the other hand:

$$\overline{I}_1 = 10j\ \text{A}$$

Substituting values

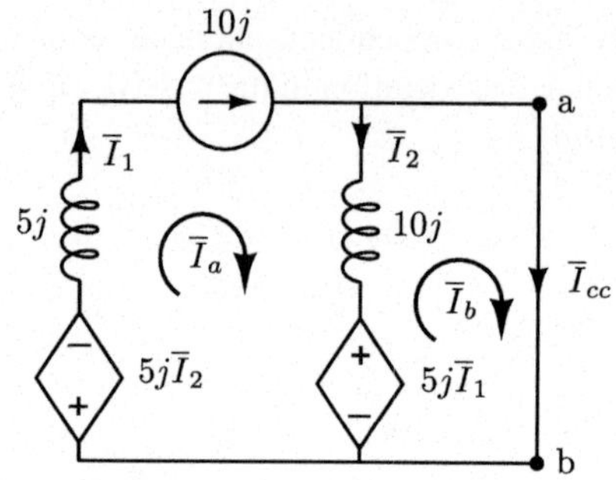

Fig. 2.175 Circuit with k open

$$\begin{pmatrix} 1 & 0 \\ -10j & 10j \end{pmatrix} \begin{pmatrix} \overline{I}_a \\ \overline{I}_b \end{pmatrix} = \begin{pmatrix} 10j \\ 5j \cdot 10j \end{pmatrix}$$

and solving yields the mesh currents:

$$\overline{I}_a = 10j \text{ A} \ ; \ \overline{I}_b = 15j \text{ A}$$

Consequently, the short-circuit current is as follows:

$$\overline{I}_{cc} = \overline{I}_b = 15j \text{ A}.$$

2.50 Maximum Power Transfer

In the circuit of Fig. 2.176 determine the resistance R_L that, once connected between terminals a and b, absorbs the maximum power. Also, calculate the value of said power.

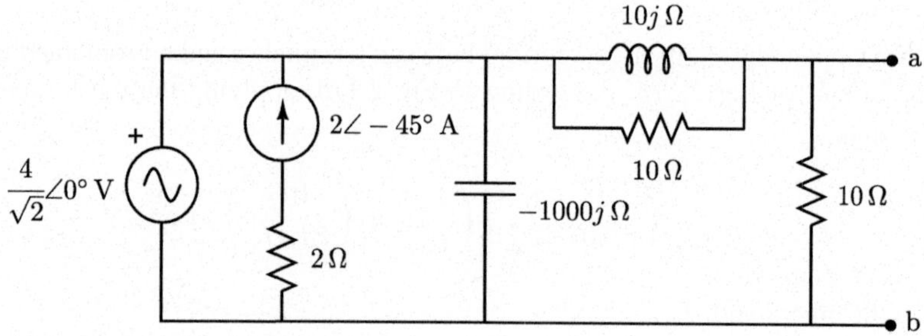

Fig. 2.176 Circuit considered in the problem under study

Solution To calculate the resistance R_L that absorbs the maximum power and the value of said power, it is necessary to obtain the Thévenin equivalent from terminals a and b.

First, the open-circuit voltage between terminals a and b, Thévenin voltage, will be calculated. To do this, the circuit shown in Fig. 2.177 will be solved. It can be observed that the current source in series with the resistor and the capacitor have been eliminated as they are connected in parallel with a voltage source.

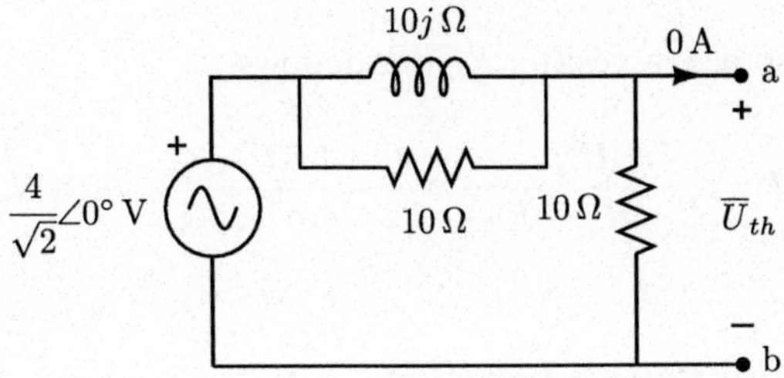

Fig. 2.177 Circuit to be solved

According to Fig. 2.177, applying the voltage divider concept yields:

$$\overline{U}_{th} = \frac{4}{\sqrt{2}}\angle 0° \cdot \frac{10}{\dfrac{10 \cdot 10j}{10 + 10j} + 10} \approx 1{,}77\angle - 18{,}43° \text{ V}$$

Next, the equivalent impedance, Thévenin impedance from terminals a and b, will be calculated. To do this, the circuit of Fig. 2.176 or that of Fig. 2.177 will be made passive, as shown in Fig. 2.178.

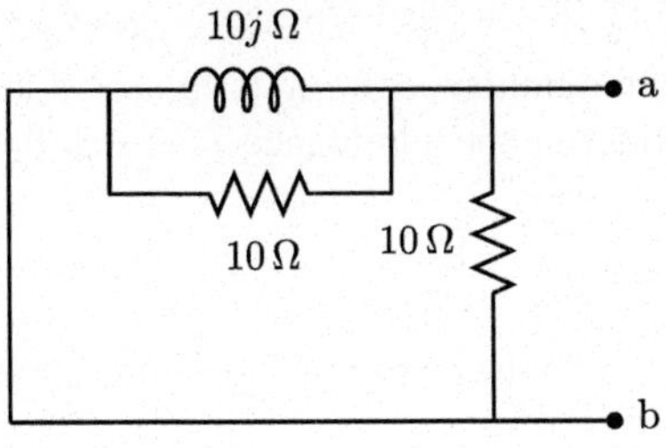

Fig. 2.178 Passive circuit

According to Fig. 2.178, the Thévenin impedance is calculated as follows:

$$\overline{Z}_{th} = 10//10//10j = \frac{1}{\frac{1}{10} + \frac{1}{10} + \frac{1}{10j}} = \underbrace{4}_{R_{th}} + \underbrace{2j}_{X_{th}\,j}\ \Omega$$

Once the equivalent impedance from terminals a and b is calculated, the resistance R_L that absorbs the maximum power is as follows:

$$R_L = \sqrt{R_{th}^2 + X_{th}^2} = \sqrt{4^2 + 2^2} \approx 4{,}47\ \Omega$$

The power absorbed by resistor R_L is as follows:

$$P = \frac{U_{th}^2}{2 \cdot (R_L + R_{th})} = \frac{1{,}77^2}{2 \cdot (4{,}47 + 4)} \approx 0{,}18\ \mathrm{W}.$$

2.51 Maximum Power Transfer

In the circuit of Fig. 2.179, $\overline{Y}$ is the admittance that absorbs maximum power from the rest of the circuit. Knowing that this power is 8 W, determine the reading of the voltmeter.

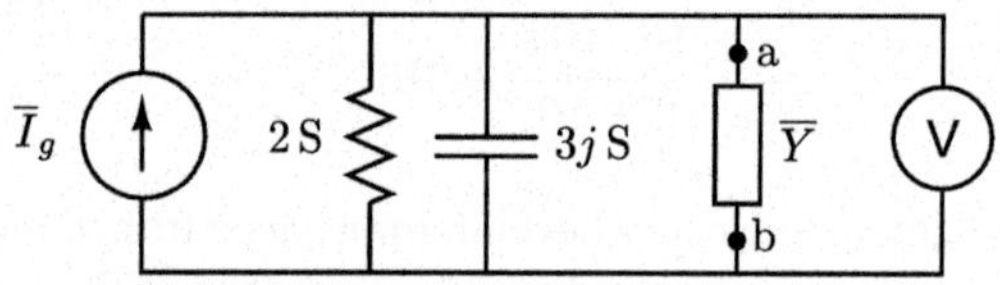

Fig. 2.179 Circuit considered in the problem under study

Solution The equivalent admittance from terminals a and b of the circuit in Fig. 2.179, without considering the admittance $\overline{Y}$, is as follows:

$$\overline{Y}_{eq} = 2 + 3j\ \mathrm{S}$$

If the admittance $\overline{Y}$ is the one absorbing maximum power from the rest of the circuit, this implies the following:

$$\overline{Y} = \overline{Y}_{eq}^{*} = \underbrace{2}_{G_{eq}} - 3j\ \mathrm{S}$$

On the other hand, the power absorbed by the admittance $\overline{Y}$ is as follows:

$$P = \frac{I_g^2}{4 \cdot G_{eq}} = \frac{I_g^2}{4 \cdot 2} = \frac{I_g^2}{8}$$

According to the statement, this power is 8 W, therefore:

$$P = \frac{I_g^2}{8} = 8\,\text{W} \Rightarrow I_g = 8\,\text{A}$$

In Fig. 2.180, the circuit with the connected admittance $\overline{Y}$ and its subsequent simplification are shown.

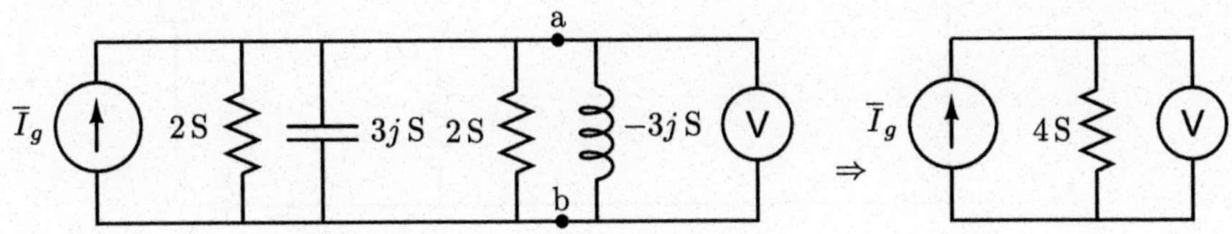

Fig. 2.180 Simplified circuit

According to Fig. 2.180, the reading of the voltmeter is as follows:

$$\text{Ⓥ} = \frac{I_g}{4} = \frac{8}{4} = 2\,\text{V}.$$

2.52 Maximum Power Transfer

In the circuit of Fig. 2.181, determine the value of $\overline{Z}$ so that it absorbs maximum power. Also, calculate said power.

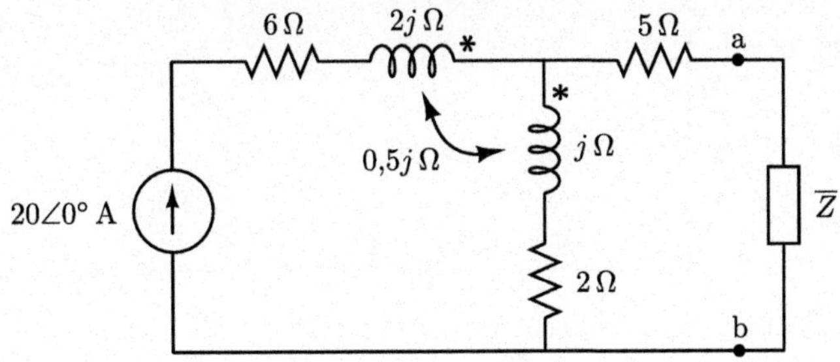

Fig. 2.181 Circuit considered in the problem under study

Solution First, the coupled coils will be replaced by their representation with dependent voltage sources in the phasor domain as shown in Fig. 2.182.

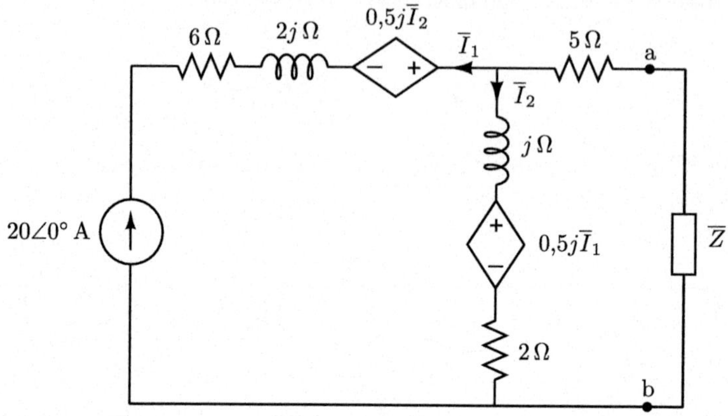

Fig. 2.182 Equivalent model of coupled coils

Next, the open-circuit voltage between terminals a and b, Thévenin voltage, will be calculated. To do this, the circuit shown in Fig. 2.183 will be solved. According to Fig. 2.183:

$$\overline{U}_{th} = j \cdot \overline{I}_2 + 0{,}5j \cdot \overline{I}_1 + 2 \cdot \overline{I}_2$$

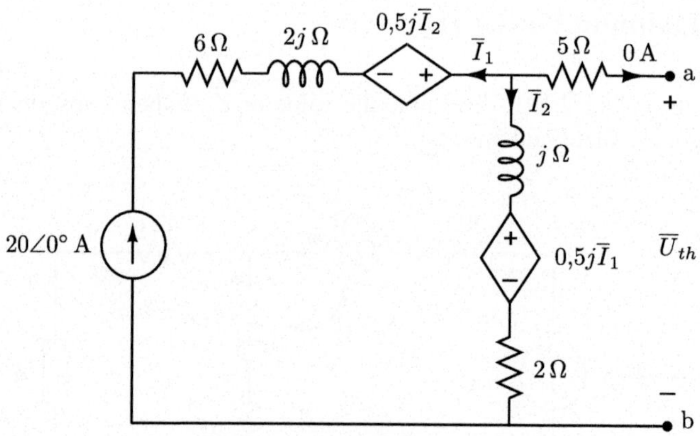

Fig. 2.183 Circuit to be solved to obtain Thévenin voltage

On the other hand:

$$\overline{I}_1 = -\overline{I}_2 = -20\angle 0^\circ \text{ A}$$

Therefore:

$$\overline{U}_{th} = j \cdot 20\angle 0° + 0{,}5j \cdot (-20\angle 0°) + 2 \cdot 20\angle 0° = 40 + 10j \approx 41{,}23\angle 14° \text{ V}$$

Next, the Norton current will be calculated, which is obtained by short-circuiting terminals a and b, as shown in Fig. 2.184.

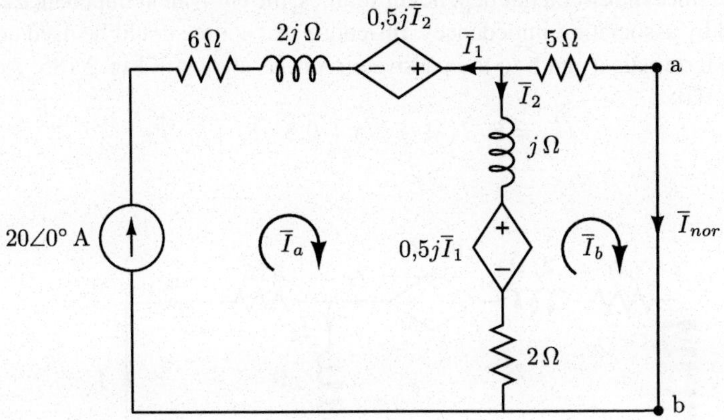

Fig. 2.184 Circuit with terminals a and b short-circuited

According to Fig. 2.184, applying the mesh analysis method:

$$\begin{pmatrix} 1 & 0 \\ -2-j & 2+5+j \end{pmatrix} \begin{pmatrix} \overline{I}_a \\ \overline{I}_b \end{pmatrix} = \begin{pmatrix} 20\angle 0° \\ 0{,}5j\overline{I}_1 \end{pmatrix}$$

On the other hand:

$$\overline{I}_1 = -\overline{I}_a \;\; ; \;\; \overline{I}_b = \overline{I}_{nor}$$

Substituting values:

$$\begin{pmatrix} 1 & 0 \\ -2-0{,}5j & 7+j \end{pmatrix} \begin{pmatrix} \overline{I}_a \\ \overline{I}_{nor} \end{pmatrix} = \begin{pmatrix} 20\angle 0° \\ -0{,}5j\overline{I}_a \end{pmatrix}$$

rearranging terms:

$$\begin{pmatrix} 1 & 0 \\ -2-0{,}5j & 7+j \end{pmatrix} \begin{pmatrix} \overline{I}_a \\ \overline{I}_{nor} \end{pmatrix} = \begin{pmatrix} 20\angle 0° \\ 0 \end{pmatrix}$$

and solving, the current $\overline{I}_{nor}$ is obtained:

$$\overline{I}_{nor} = 5{,}8 + 0{,}6j \approx 5{,}83\angle 5{,}91° \text{ A}$$

Once the Thévenin voltage and the Norton current are obtained, the equivalent impedance between terminals a and b is calculated as follows:

$$\overline{Z}_{eq} = \frac{\overline{U}_{th}}{\overline{I}_{nor}} = \frac{40+10j}{5{,}8+0{,}6j} = 7+j\,\Omega$$

Another way to calculate the Thévenin impedance is from the passive circuit. In this case, since the circuit has dependent sources, the equivalent impedance cannot be obtained by associating impedances. Instead, a test source could be used connected between terminals a and b to the passive circuit, as shown in Fig. 2.185. According to Fig. 2.185:

$$\overline{U}_x = 5\cdot\overline{I}_x + j\cdot\overline{I}_2 + 0{,}5\cdot\overline{I}_1 + 2\cdot\overline{I}_2$$

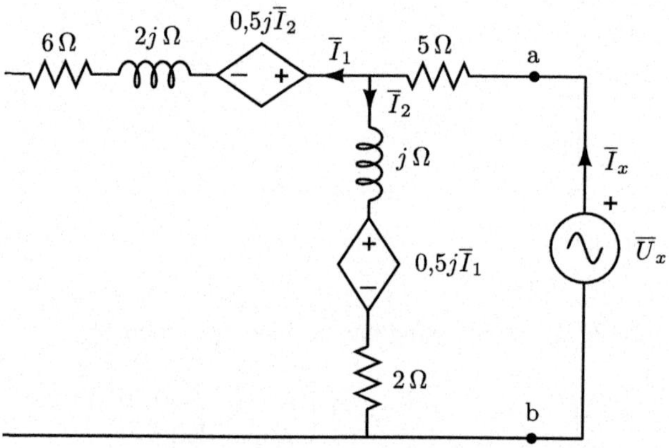

Fig. 2.185 Circuit with a test voltage source

On the other hand:

$$\overline{I}_1 = 0\,\text{A}\ ;\ \overline{I}_2 = \overline{I}_x$$

Therefore:

$$\overline{U}_x = 5\cdot\overline{I}_x + j\cdot\overline{I}_x + 2\cdot\overline{I}_x = (7+j)\cdot\overline{I}_x \ \Rightarrow\ \overline{Z}_{eq} = \frac{\overline{U}_x}{\overline{I}_x} = \underbrace{7}_{R_{eq}} + j\,\Omega$$

Once the equivalent impedance from terminals a and b is calculated, the impedance $\overline{Z}$ that absorbs maximum power is as follows:

$$\overline{Z} = \overline{Z}_{eq}^{*} = 7 - j\,\Omega$$

On the other hand, the power absorbed by said impedance will be as follows:

$$P = \frac{U_{th}^2}{4 \cdot R_{eq}} = \frac{41{,}23^2}{4 \cdot 7} \approx 60{,}71\,\text{W}.$$

2.53 Maximum Power Transfer

In the circuit of Fig. 2.186, determine the value of the admittance $\overline{Y}$ such that the power factor of the source is unity. Also, determine the maximum power that said admittance can absorb.

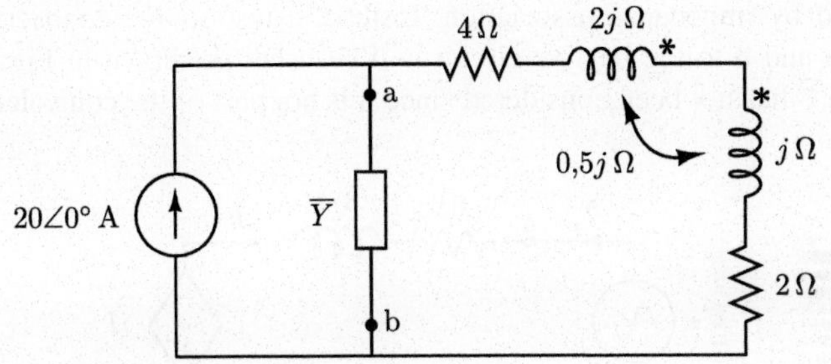

Fig. 2.186 Circuit considered in the problem under study

Solution First, the coupled coils will be replaced by their representation with dependent voltage sources in the phasor domain, as shown in Fig. 2.187. The circuit of Fig. 2.187 can be simplified by associating sources and impedances, resulting in the circuit of Fig. 2.188.

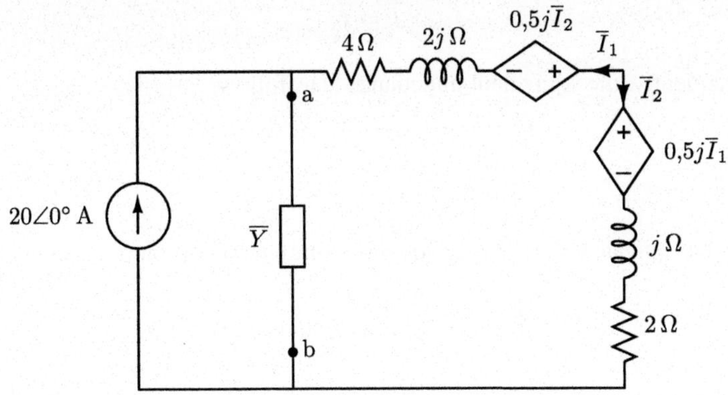

Fig. 2.187 Equivalent model of coupled coils

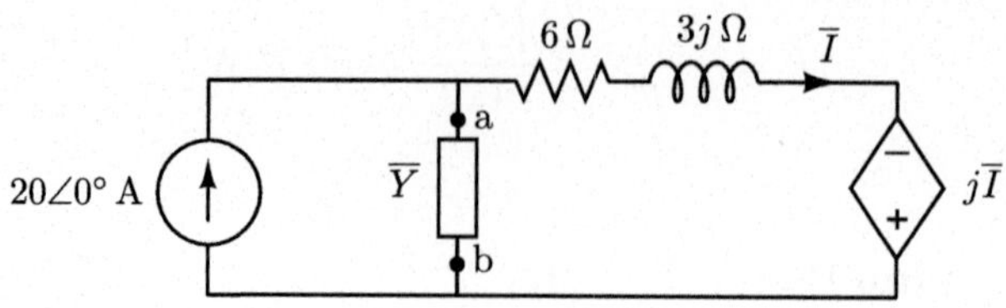

Fig. 2.188 Simplified circuit

Next, the equivalent impedance from terminals a and b will be calculated. In this case, since the circuit has a dependent source, the equivalent impedance cannot be obtained by impedance association. Instead, a test source connected between terminals a and b to the passive circuit will be used, as shown in Fig. 2.189. The admittance $\overline{Y}$ has not been considered since it is not part of the equivalent.

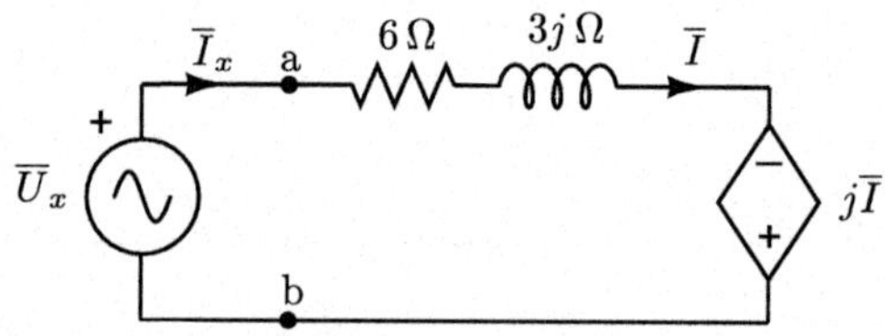

Fig. 2.189 Circuit with a test voltage source

According to Fig. 2.189:

$$\overline{U}_x = (6+3j)\cdot\overline{I} - j\overline{I} = (6+2j)\cdot\overline{I}$$

On the other hand:

$$\overline{I} = \overline{I}_x$$

Consequently, the equivalent impedance is as follows:

$$\overline{Z}_{eq} = \frac{\overline{U}_x}{\overline{I}_x} = 6+2j\ \Omega$$

Once the equivalent impedance is obtained, the equivalent admittance is calculated as follows:

$$\overline{Y}_{eq} = \frac{1}{\overline{Z}_{eq}} = \frac{1}{6+2j} = 0{,}15 - 0{,}05j\ \mathrm{S}$$

For the source power factor to be unity once the admittance $\overline{Y}$ is connected between terminals a and b, it is necessary for the imaginary part of $\overline{Y}$ to be equal and opposite in sign to the imaginary part of $\overline{Y}_{eq}$. Thus, the total admittance, obtained as the sum of $\overline{Y}$ and $\overline{Y}_{eq}$, only has a real part. Consequently, the admittance $\overline{Y}$ is as follows:

$$\overline{Y} = G + 0{,}05j \text{ S}$$

It can be observed that G can take any value. However, for $\overline{Y}$ to absorb the maximum power, then:

$$\overline{Y} = \overline{Y}^*_{eq} = 0{,}15 + 0{,}05j \text{ S}$$

In this case, the power absorbed by $\overline{Y}$ is as follows:

$$P = \frac{20^2}{4 \cdot 0{,}15} \approx 666{,}67 \text{ W}.$$

2.54 Maximum Power Transfer

In the circuit of Fig. 2.190, it is requested to:

1. Calculate the Thévenin and Norton equivalents from terminals a and b.
2. Determine the resistance R that, once connected between a and b, absorbs the maximum power.
3. Determine the impedance $\overline{Z}$ that, once connected between a and b, absorbs the maximum power.

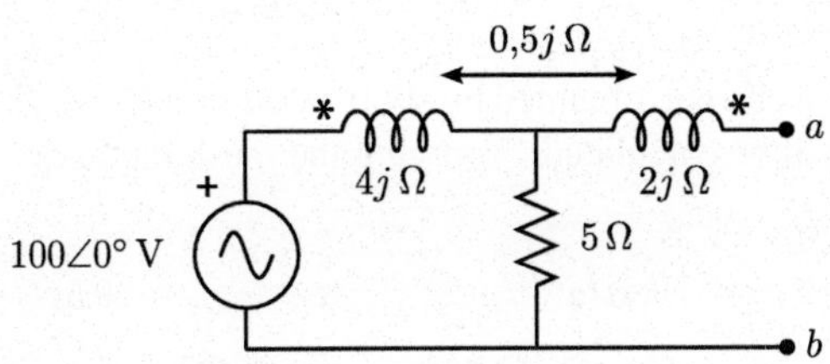

Fig. 2.190 Circuit considered in the problem under study

Solution First, the coupled coils will be replaced by their representation with dependent voltage sources in the phasor domain, as shown in Fig. 2.191.

1. Next, the open-circuit voltage between terminals a and b, Thévenin voltage, will be calculated. For this, the circuit shown in Fig. 2.192 will be solved.
   According to Fig. 2.192, considering that $\overline{I}_2 = 0$:

   $$\overline{U}_{th} = 0{,}5j \cdot \overline{I}_1 + 5 \cdot \overline{I}_1$$

   On the other hand:

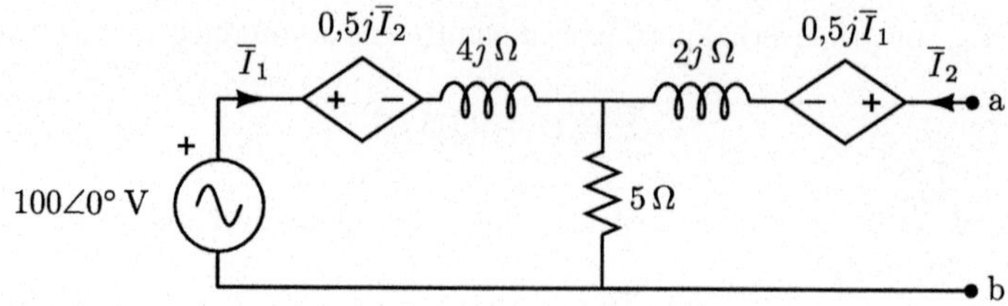

Fig. 2.191 Equivalent model of coupled coils

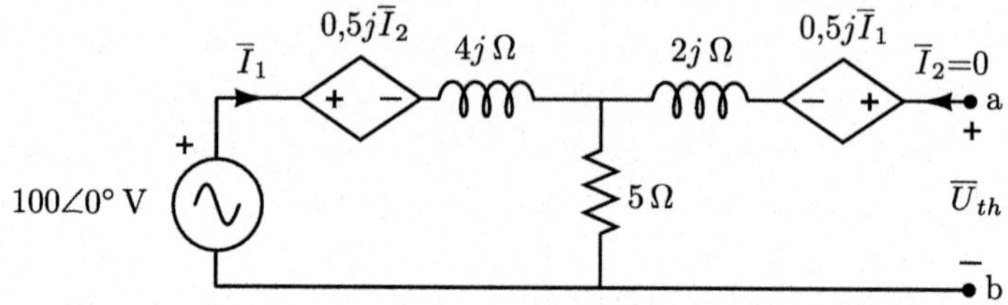

Fig. 2.192 Circuit to be solved to obtain Thévenin voltage

$$100\angle 0^\circ = 4j \cdot \overline{I}_1 + 5 \cdot \overline{I}_1 \;\Rightarrow\; \overline{I}_1 = \frac{100\angle 0^\circ}{5+4j}\text{ A}$$

Consequently:

$$\overline{U}_{th} = (5+0{,}5j) \cdot \overline{I}_1 = (5+0{,}5j) \cdot \frac{100\angle 0^\circ}{5+4j} \approx 78{,}47\angle -32{,}95^\circ\text{ V}$$

Then, the Norton current, obtained by short-circuiting terminals a and b, as shown in Fig. 2.193, will be calculated. Applying the mesh method:

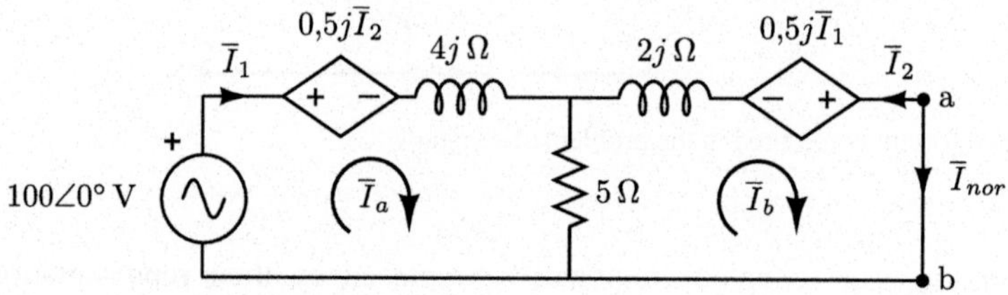

Fig. 2.193 Circuit with terminals a and b short-circuited

$$\begin{pmatrix} 5+4j & -5 \\ -5 & 5+2j \end{pmatrix} \begin{pmatrix} \overline{I}_a \\ \overline{I}_b \end{pmatrix} = \begin{pmatrix} 100\angle 0^\circ - 0{,}5j\overline{I}_2 \\ 0{,}5j\overline{I}_1 \end{pmatrix}$$

On the other hand:

$$\overline{I}_1 = \overline{I}_a \ ; \ \overline{I}_2 = -\overline{I}_b = -\overline{I}_{nor}$$

Substituting values:

$$\begin{pmatrix} 5+4j & -5 \\ -5 & 5+2j \end{pmatrix} \begin{pmatrix} \overline{I}_a \\ \overline{I}_{nor} \end{pmatrix} = \begin{pmatrix} 100\angle 0^\circ + 0{,}5j\overline{I}_{nor} \\ 0{,}5j\overline{I}_a \end{pmatrix}$$

rearranging terms:

$$\begin{pmatrix} 5+4j & -5-0{,}5j \\ -5-0{,}5j & 5+2j \end{pmatrix} \begin{pmatrix} \overline{I}_a \\ \overline{I}_{nor} \end{pmatrix} = \begin{pmatrix} 100\angle 0^\circ \\ 0 \end{pmatrix}$$

and solving yields the Norton current $\overline{I}_{nor}$:

$$\overline{I}_{nor} \approx 19{,}2\angle - 101{,}51^\circ \text{ A}$$

Once the Thévenin voltage and the Norton current are obtained, the equivalent impedance between terminals a and b is calculated as follows:

$$\overline{Z}_{eq} = \frac{\overline{U}_{th}}{\overline{I}_{nor}} = \frac{78{,}47\angle - 32{,}95^\circ}{19{,}2\angle - 101{,}51^\circ} \approx 1{,}49 + 3{,}8j\ \Omega$$

Another way to calculate the equivalent impedance is from the passive circuit. In this case, since the circuit has dependent sources, the equivalent impedance cannot be obtained by impedance association. Instead, a test source connected between terminals a and b to the passive circuit, as shown in Fig. 2.194, could be used.

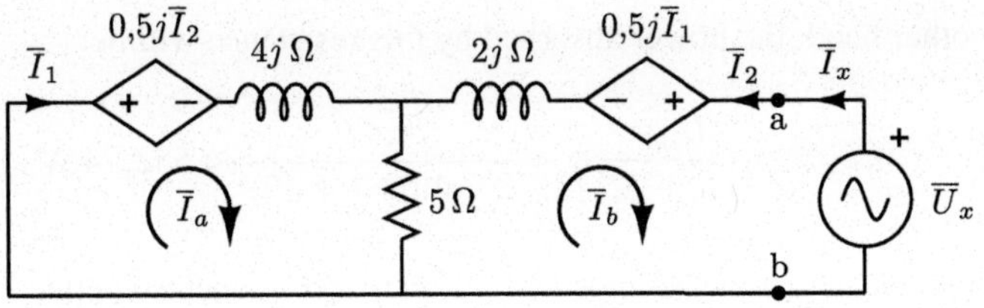

Fig. 2.194 Circuit with a test voltage source

According to Fig. 2.194, applying the mesh method:

$$\begin{pmatrix} 5+4j & -5 \\ -5 & 5+2j \end{pmatrix} \begin{pmatrix} \overline{I}_a \\ \overline{I}_b \end{pmatrix} = \begin{pmatrix} -0{,}5j\overline{I}_2 \\ 0{,}5j\overline{I}_1 - \overline{U}_x \end{pmatrix}$$

On the other hand:

$$\overline{I}_1 = \overline{I}_a \ ; \ \overline{I}_2 = -\overline{I}_b = -\overline{I}_x$$

Substituting values:

$$\begin{pmatrix} 5+4j & -5 \\ -5 & 5+2j \end{pmatrix} \begin{pmatrix} \overline{I}_a \\ -\overline{I}_x \end{pmatrix} = \begin{pmatrix} 0 \\ -\overline{U}_x \end{pmatrix}$$

and solving yields the relationship between $\overline{U}_x$ and $\overline{I}_x$:

$$\overline{Z}_{eq} = \frac{\overline{U}_x}{\overline{I}_x} = (5+0{,}5j) \cdot \left[\frac{5+2j}{5+0{,}5j} - \frac{5+0{,}5j}{5+4j} \right] \approx \underbrace{1{,}49}_{R_{eq}} + \underbrace{3{,}8}_{X_{eq}} \, j \, \Omega$$

2. Once the equivalent impedance from terminals a and b is calculated, then the resistance R that, once connected between terminals a and b, absorbs the maximum power, is calculated considering the circuit of Fig. 2.195.

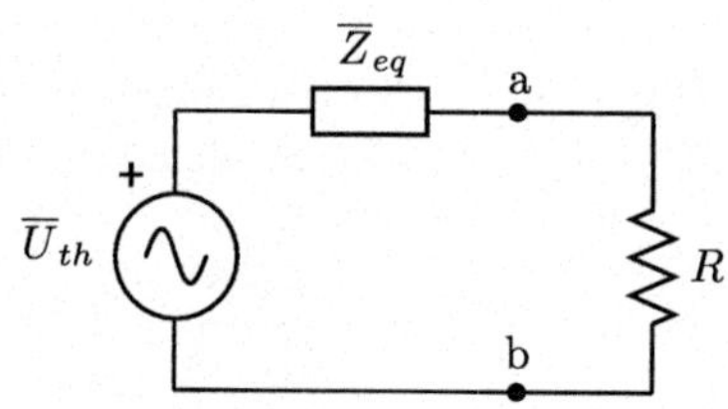

Fig. 2.195 Circuit for the calculation R

The value of resistance R is as follows:

$$R^2 = R_{eq}^2 + X_{eq}^2 = 1{,}49^2 + 3{,}8^2 \approx 16{,}66 \, \Omega^2 \;\Rightarrow\; R \approx 4{,}08 \, \Omega$$

On the other hand, the power absorbed by this resistance will be:

$$P_R = \frac{U_{th}^2}{2 \cdot (R + R_{eq})} = \frac{78{,}47^2}{2 \cdot (4{,}08 + 1{,}49)} \approx 552{,}74 \, \text{W}$$

3. In the case of calculating the impedance $\overline{Z}$ that, once connected between terminals a and b, absorbs the maximum power, the circuit of Fig. 2.196 must be considered. The value of impedance $\overline{Z}$ is as follows:

$$\overline{Z} = \overline{Z}_{eq}^* = 1{,}49 - 3{,}8j \, \Omega$$

On the other hand, the power absorbed by this impedance will be:

$$P_Z = \frac{U_{th}^2}{4 \cdot R_{eq}} = \frac{78{,}47^2}{4 \cdot 1{,}49} \approx 1\,033{,}14 \, \text{W}$$

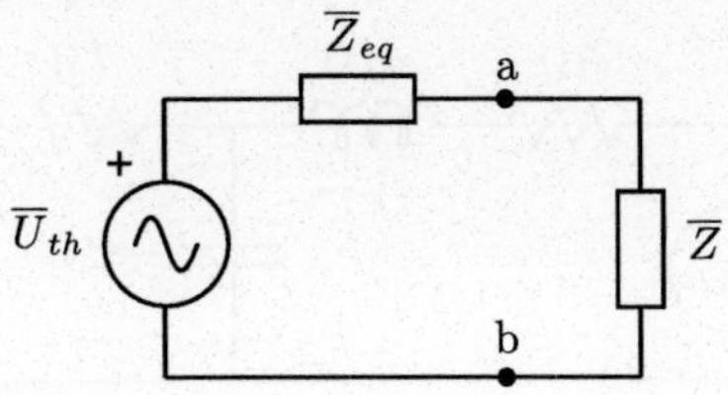

Fig. 2.196 Circuit for the calculation $\overline{Z}$

As expected, it can be observed that the power absorbed by impedance $\overline{Z}$ is greater than the power absorbed by resistance R.

2.55 Maximum Power Transfer

The resistance R of the circuit in Fig. 2.197 is adjusted until the power absorbed reaches its maximum. Determine the value of R and the power absorbed by said resistance.

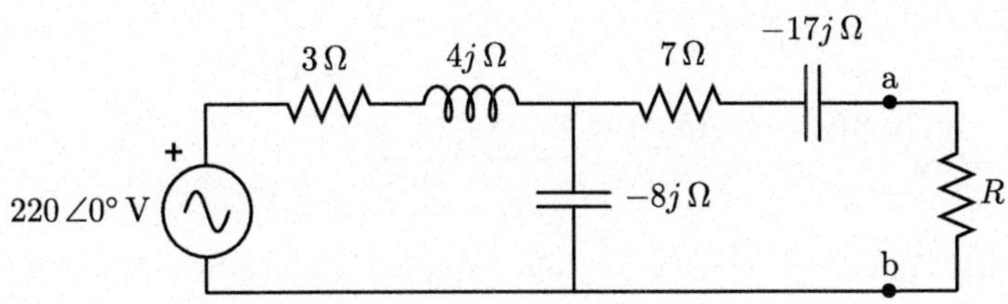

Fig. 2.197 Circuit considered in the problem under study

Solution To calculate the resistance R that absorbs the maximum power and the value of that power, it is necessary to obtain the Thévenin equivalent from terminals a and b.

First, the open-circuit voltage between terminals a and b, Thévenin voltage, will be calculated. To do this, the circuit shown in Fig. 2.198 will be solved. It can be noted that the resistance R has not been considered since it is not part of the equivalent.

According to Fig. 2.198, applying the voltage divider concept:

$$\overline{U}_{th} = 220\angle 0^\circ \cdot \frac{-8j}{3+4j-8j} \approx 352\angle -36{,}87^\circ \text{ V}$$

Next, the equivalent impedance, Thévenin impedance from terminals a and b, will be calculated. To do this, the sources of the circuit in Fig. 2.198 will be nullified, as shown in Fig. 2.199.

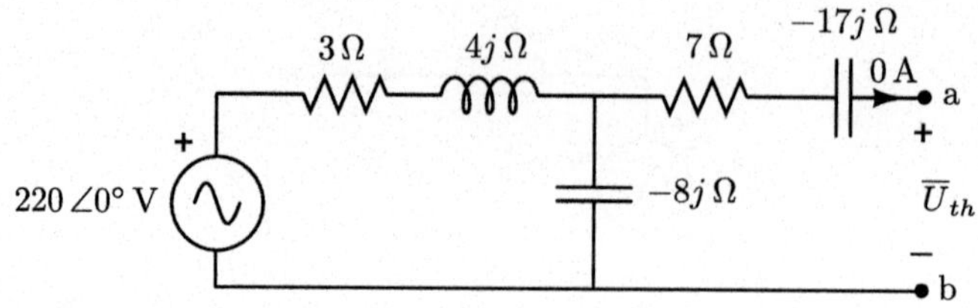

Fig. 2.198 Circuit to be solved to obtain Thévenin voltage

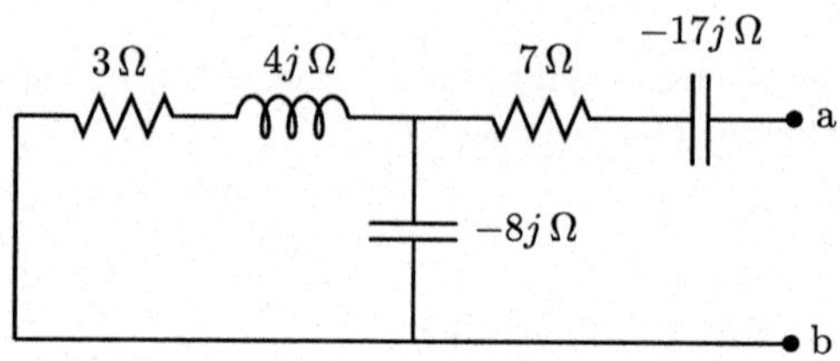

Fig. 2.199 Passive circuit

According to Fig. 2.199, the Thévenin impedance is calculated as follows:

$$\overline{Z}_{th} = [(3+4j)//(-8j)] + 7 - 17j = \frac{(3+4j)\cdot(-8j)}{3+4j-8j} + 7 - 17j$$
$$\approx 20{,}81\angle -45{,}15^\circ\ \Omega \approx \underbrace{14{,}68}_{R_{th}} \underbrace{-14{,}76}_{X_{th}}\, j\ \Omega$$

Once the equivalent impedance from terminals a and b is calculated, then the resistance R that absorbs the maximum power is as follows:

$$R = \sqrt{R_{th}^2 + X_{th}^2} = \sqrt{14{,}68^2 + (-14{,}76)^2} \approx 20{,}81\ \Omega$$

The power absorbed by resistance R is as follows:

$$P = \frac{U_{th}^2}{2\cdot(R+R_{th})} = \frac{352^2}{2\cdot(20{,}81+14{,}68)} \approx 1\,745{,}6\,\mathrm{W}.$$

2.56 Maximum Power Transfer

In the circuit of Fig. 2.200, determine the resistance R_L that, once connected between terminals a and b, absorbs the maximum power. Also, calculate said power.

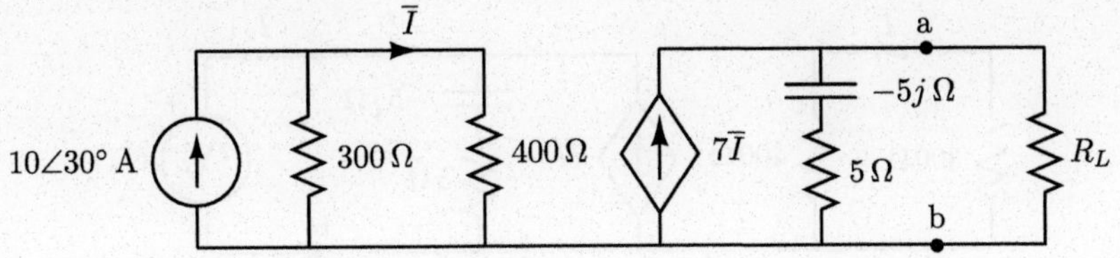

Fig. 2.200 Circuit considered in the problem under study

Solution To calculate the resistance R_L that, once connected between terminals a and b, absorbs the maximum power and the value of that power, it is necessary to obtain the Thévenin equivalent from terminals a and b.

First, the open-circuit voltage between terminals a and b, Thévenin voltage, will be calculated. To do this, the circuit shown in Fig. 2.201 will be solved. It can be noted that the resistance R_L has not been considered since it is not part of the equivalent.

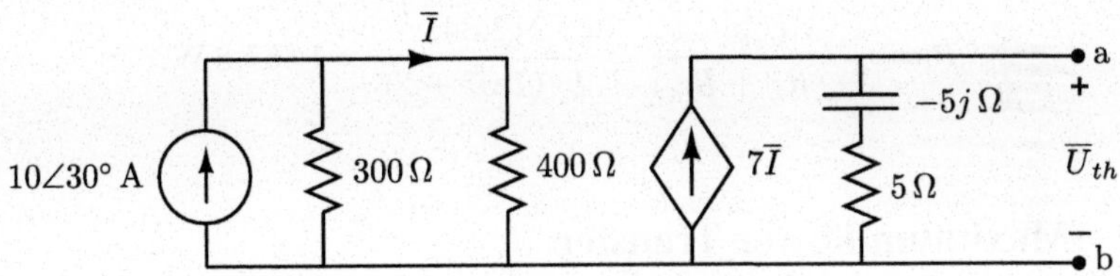

Fig. 2.201 Circuit to be solved to obtain Thévenin voltage

According to Fig. 2.201:

$$\overline{U}_{th} = 7\overline{I} \cdot (5 - 5j)$$

On the other hand, applying the concept of current divider:

$$\overline{I} = 10\angle 30° \cdot \frac{1/400}{1/300 + 1/400} \approx 4{,}29\angle 30° \text{ V}$$

Consequently:

$$\overline{U}_{th} = 7\overline{I} \cdot (5 - 5j) = 7 \cdot 4{,}29\angle 30° \cdot (5 - 5j) \approx 212{,}34\angle - 15° \text{ V}$$

Next, the equivalent impedance, Thévenin impedance from terminals a and b, will be calculated. To do this, the circuit in Fig. 2.201 will be made passive, resulting in the circuit in Fig. 2.202, to which a test source of value $\overline{U}_x$ has been connected.

According to Fig. 2.202, it can be verified that $\overline{I} = 0$ A. Therefore:

$$\overline{U}_x = \overline{I}_x \cdot (5 - 5j) \;\Rightarrow\; \overline{Z}_{th} = \frac{\overline{U}_x}{\overline{I}_x} = \underbrace{5}_{R_{th}} \underbrace{-5}_{X_{th}} j\,\Omega$$

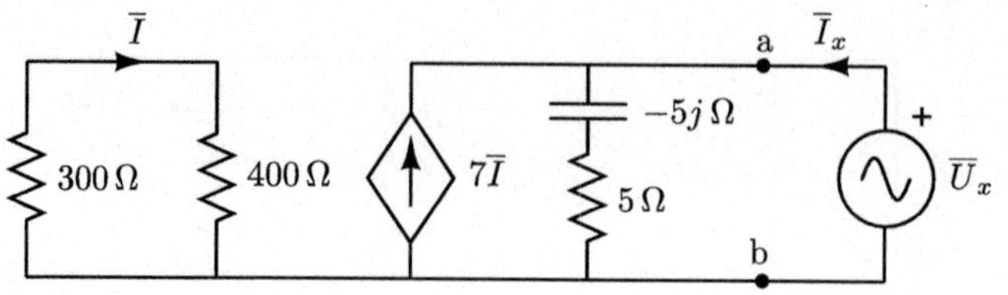

Fig. 2.202 Passive circuit

Once the equivalent impedance from terminals a and b is calculated, the resistance R_L that absorbs the maximum power is as follows:

$$R_L = \sqrt{R_{th}^2 + X_{th}^2} = \sqrt{5^2 + (-5)^2} \approx 7{,}07\,\Omega$$

The power absorbed by resistance R_L is as follows:

$$P = \frac{U_{th}^2}{2 \cdot (R_L + R_{th})} = \frac{212{,}34^2}{2 \cdot (7{,}07 + 5)} \approx 1\,867{,}8\,\text{W}.$$

2.57 Maximum Power Transfer

In the circuit of Fig. 2.203, calculate the value of $\overline{Z}$ that, once connected between terminals a and b, absorbs the maximum power. Also, calculate said power.

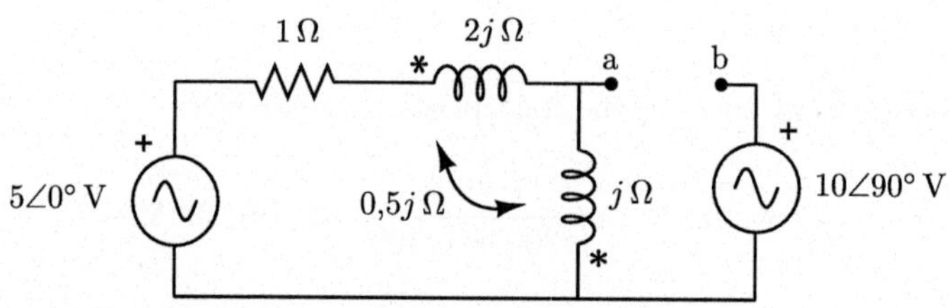

Fig. 2.203 Circuit considered in the problem under study

Solution First, the coupled coils will be replaced by their representation with dependent voltage sources in the phasor domain, as shown in Fig. 2.204.

Next, the open-circuit voltage between terminals a and b, Thévenin voltage, will be calculated. To do this, the circuit shown in Fig. 2.205 will be solved.

According to Fig. 2.205:

$$\overline{U}_{th} = -j \cdot \overline{I}_2 - 0{,}5j \cdot \overline{I}_1 - 10\angle 90^\circ$$

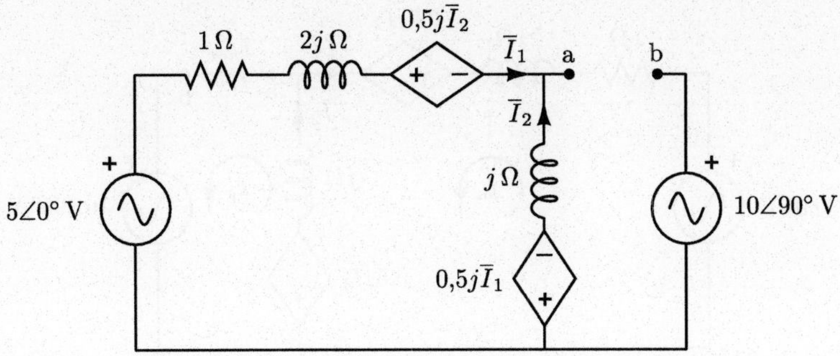

Fig. 2.204 Equivalent model of coupled coils

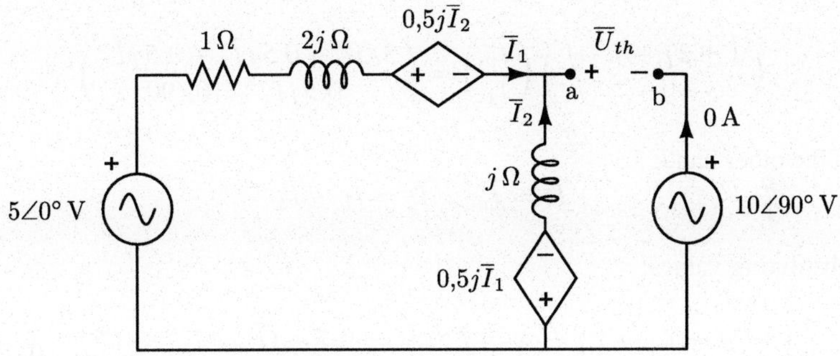

Fig. 2.205 Circuit to be solved to obtain Thévenin voltage

On the other hand:

$$5\angle 0^\circ = (1+2j)\cdot \overline{I}_1 + 0{,}5j\cdot \overline{I}_2 - j\cdot \overline{I}_2 - 0{,}5j\cdot \overline{I}_1$$

Taking into account that $\overline{I}_1 = -\overline{I}_2$, then:

$$5\angle 0^\circ = (1+2j)\cdot \overline{I}_1 - 0{,}5j\cdot \overline{I}_1 + j\cdot \overline{I}_1 - 0{,}5j\cdot \overline{I}_1 \;\Rightarrow\; \overline{I}_1 = \frac{5\angle 0^\circ}{1+2j}\,\text{A}$$

Thus, the Thévenin voltage is:

$$\overline{U}_{th} = j\cdot \overline{I}_1 - 0{,}5j\cdot \overline{I}_1 - 10\angle 90^\circ = 0{,}5j\cdot \frac{5\angle 0^\circ}{1+2j} - 10\angle 90^\circ \approx 9{,}55\angle -84^\circ\,\text{V}$$

Next, the Norton current will be calculated, which is obtained by short-circuiting terminals a and b, as shown in Fig. 2.206.

According to Fig. 2.206, applying the mesh analysis method:

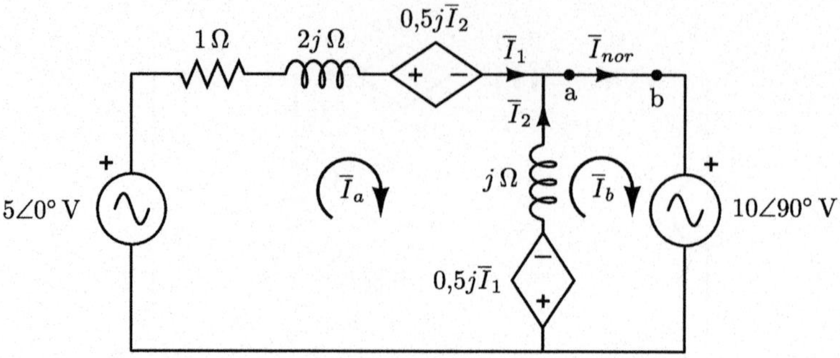

Fig. 2.206 Circuit with terminals a and b short-circuited

$$\begin{pmatrix} 1+2j+j & -j \\ -j & j \end{pmatrix} \begin{pmatrix} \overline{I}_a \\ \overline{I}_b \end{pmatrix} = \begin{pmatrix} 5\angle 0^\circ - 0{,}5j\overline{I}_2 + 0{,}5j\overline{I}_1 \\ -0{,}5j\overline{I}_1 - 10\angle 90^\circ \end{pmatrix}$$

On the other hand:

$$\overline{I}_1 = \overline{I}_a \ ; \ \overline{I}_2 = \overline{I}_b - \overline{I}_a$$

Substituting values:

$$\begin{pmatrix} 1+2j+j & -j \\ -j & j \end{pmatrix} \begin{pmatrix} \overline{I}_a \\ \overline{I}_b \end{pmatrix} = \begin{pmatrix} 5\angle 0^\circ - 0{,}5j\cdot(\overline{I}_b - \overline{I}_a) + 0{,}5j\overline{I}_a \\ -0{,}5j\overline{I}_a - 10\angle 90^\circ \end{pmatrix}$$

rearranging terms:

$$\begin{pmatrix} 1+2j & -0{,}5j \\ -0{,}5j & j \end{pmatrix} \begin{pmatrix} \overline{I}_a \\ \overline{I}_b \end{pmatrix} = \begin{pmatrix} 5\angle 0^\circ \\ -10\angle 90^\circ \end{pmatrix}$$

and solving yields the current $\overline{I}_b$, which corresponds to the Norton current, $\overline{I}_{nor}$:

$$\overline{I}_{nor} = 10{,}6\angle -170{,}8^\circ \text{ A}$$

Once the Thévenin voltage and the Norton current are obtained, the equivalent impedance between terminals a and b, Thévenin impedance, is calculated as follows:

$$\overline{Z}_{th} = \frac{\overline{U}_{th}}{\overline{I}_{nor}} = \frac{9{,}55\angle -84^\circ}{10{,}6\angle -170{,}8^\circ} \approx 0{,}9\angle 96{,}8^\circ \approx 0{,}05 + 0{,}9\,\Omega$$

Another way to calculate the Thévenin impedance is from the passive circuit. In this case, since the circuit has dependent sources, the Thévenin impedance cannot be obtained by impedance association. Instead, a test source connected between terminals a and b to the passive circuit could be used as shown in Fig. 2.207.

According to Fig. 2.207, applying the mesh analysis method:

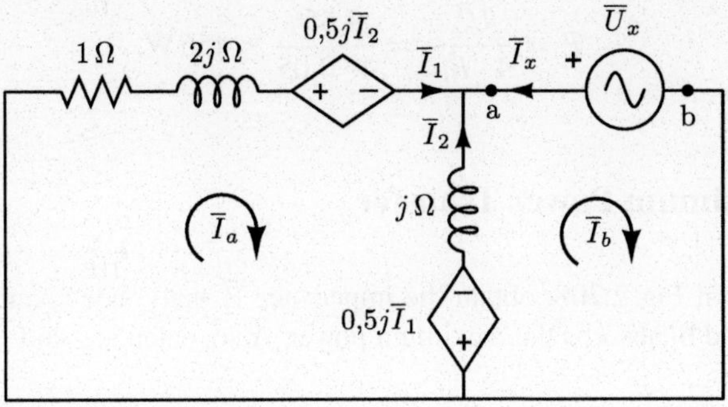

Fig. 2.207 Circuit with a test voltage source

$$\begin{pmatrix} 1+2j+j & -j \\ -j & j \end{pmatrix}\begin{pmatrix} \overline{I}_a \\ \overline{I}_b \end{pmatrix} = \begin{pmatrix} -0{,}5j\overline{I}_2 + 0{,}5j\overline{I}_1 \\ -0{,}5j\overline{I}_1 - \overline{U}_x \end{pmatrix}$$

On the other hand:

$$\overline{I}_1 = \overline{I}_a \ ; \ \overline{I}_2 = \overline{I}_b - \overline{I}_a$$

Substituting values:

$$\begin{pmatrix} 1+2j+j & -j \\ -j & j \end{pmatrix}\begin{pmatrix} \overline{I}_a \\ \overline{I}_b \end{pmatrix} = \begin{pmatrix} -0{,}5j\cdot(\overline{I}_b - \overline{I}_a) + 0{,}5j\overline{I}_a \\ -0{,}5j\overline{I}_a - \overline{U}_x \end{pmatrix}$$

rearranging terms:

$$\begin{pmatrix} 1+2j & -0{,}5j \\ -0{,}5j & j \end{pmatrix}\begin{pmatrix} \overline{I}_a \\ \overline{I}_b \end{pmatrix} = \begin{pmatrix} 0 \\ -\overline{U}_x \end{pmatrix}$$

and solving yields the current $\overline{I}_b$:

$$\overline{I}_b \approx 1{,}11\angle 93{,}18^\circ \cdot \overline{U}_x$$

Considering that $\overline{I}_b = -\overline{I}_x$, then the relationship between $\overline{U}_x$ and $\overline{I}_x$ is as follows:

$$\overline{Z}_{th} = \frac{\overline{U}_x}{\overline{I}_x} = \frac{-1}{1{,}11\angle 93{,}18^\circ} \approx 0{,}9\angle 86{,}82^\circ \approx \underbrace{0{,}05}_{R_{th}} + 0{,}9j\ \Omega$$

The value of the impedance $\overline{Z}$ that, once connected between terminals a and b, absorbs the maximum power is as follows:

$$\overline{Z} = \overline{Z}^*_{th} = 0{,}05 - 0{,}9j\ \Omega$$

On the other hand, the power absorbed by this impedance is as follows:

$$P = \frac{U_{th}^2}{4 \cdot R_{th}} = \frac{9{,}55^2}{4 \cdot 0{,}05} \approx 456\,\mathrm{W}.$$

2.58 Maximum Power Transfer

In the circuit of Fig. 2.208, obtain the impedance $\overline{Z}$ that, once connected between terminals a and b, absorbs the maximum power. Also, calculate said power.

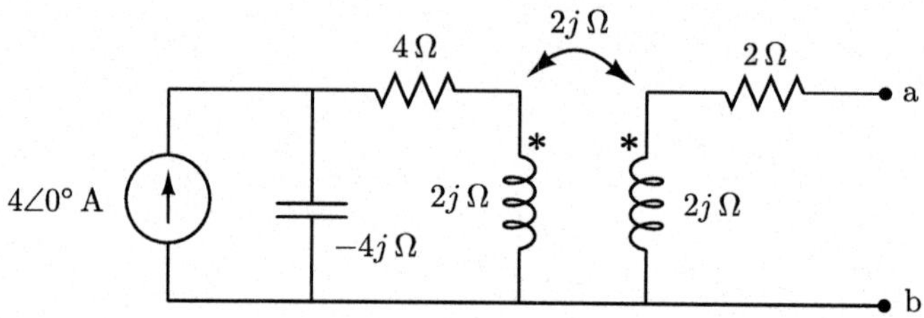

Fig. 2.208 Circuit considered in the problem under study

Solution First, the coupled coils will be replaced by their representation with dependent voltage sources in the phasor domain, as shown in Fig. 2.209.

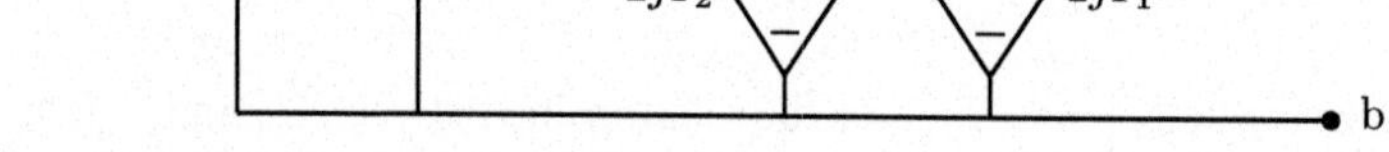

Fig. 2.209 Equivalent model of coupled coils

Next, the open-circuit voltage between terminals a and b, Thévenin voltage, will be calculated. To do this, the circuit shown in Fig. 2.210 will be solved.

According to Fig. 2.210, it is easy to verify that $\overline{I}_2 = 0\,\mathrm{A}$. This makes the dependent voltage source on the left part of the circuit zero, and the concept of current divider can be applied to calculate the current $\overline{I}_1$:

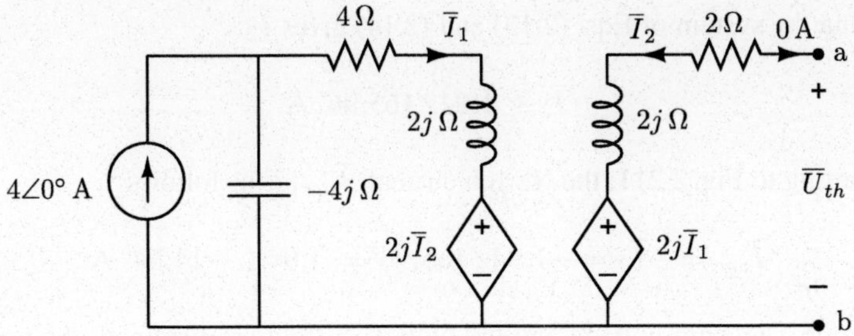

Fig. 2.210 Circuit to be solved to obtain Thévenin voltage

$$\overline{I}_1 = 4\angle 0^\circ \cdot \frac{\dfrac{1}{4+2j}}{\dfrac{1}{4+2j}+\dfrac{1}{-4j}} = 1{,}6 - 3{,}2j\ \Omega$$

Once $\overline{I}_1$ is obtained, then the Thévenin voltage is as follows:

$$\overline{U}_{th} = 2j \cdot \overline{I}_1 = 2j \cdot (1{,}6 - 3{,}2j) = 6{,}4 + 3{,}2j \approx 7{,}16\angle 26{,}57^\circ\ \text{V}$$

Next, the Norton current will be calculated, which is obtained by short-circuiting terminals a and b, as shown in Fig. 2.211. Applying Kirchhoff's voltage law in the right part of the circuit yields:

$$(2+2j) \cdot \overline{I}_2 + 2j \cdot \overline{I}_1 = 0 \tag{2.13}$$

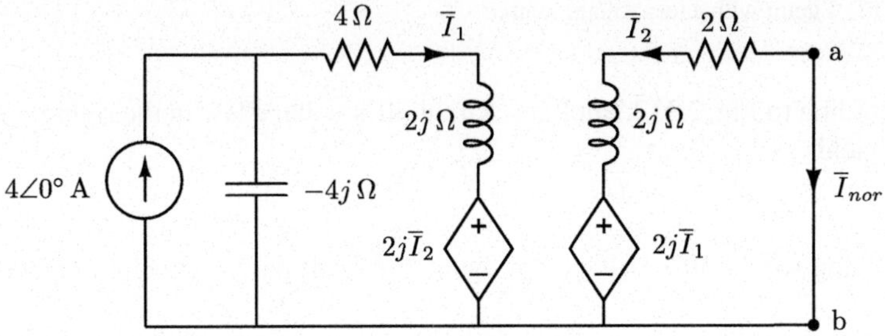

Fig. 2.211 Circuit with terminals a and b short-circuited

On the other hand, applying Kirchhoff's voltage law in the left part of the circuit (the real current source has been previously replaced by its equivalent voltage source), we obtain:

$$(-4j + 4 + 2j) \cdot \overline{I}_1 + 2j\overline{I}_2 + 16j = 0 \tag{2.14}$$

Solving the system of Eqs. (2.13) and (2.14) gives $\overline{I}_2$:

$$\overline{I}_2 \approx 1{,}94\angle 165{,}96^\circ \text{ A}$$

According to Fig. 2.211, the Norton current, $\overline{I}_{nor}$, is as follows:

$$\overline{I}_{nor} = -\overline{I}_2 = -1{,}94\angle 165{,}96^\circ = 1{,}94\angle -14{,}04^\circ \text{ A}$$

Once the Thévenin voltage and the Norton current are obtained, the equivalent impedance between terminals a and b, Thévenin impedance, is calculated as follows:

$$\overline{Z}_{th} = \frac{\overline{U}_{th}}{\overline{I}_{nor}} = \frac{7{,}16\angle 26{,}57^\circ}{1{,}94\angle -14{,}04^\circ} \approx 3{,}69\angle 40{,}6^\circ \ \Omega$$

Another way to calculate the Thévenin impedance is from the passive circuit. In this case, since the circuit has dependent sources, the Thévenin impedance cannot be obtained by impedance association. Instead, a test source connected between terminals a and b to the passive circuit could be used, as shown in Fig. 2.212.

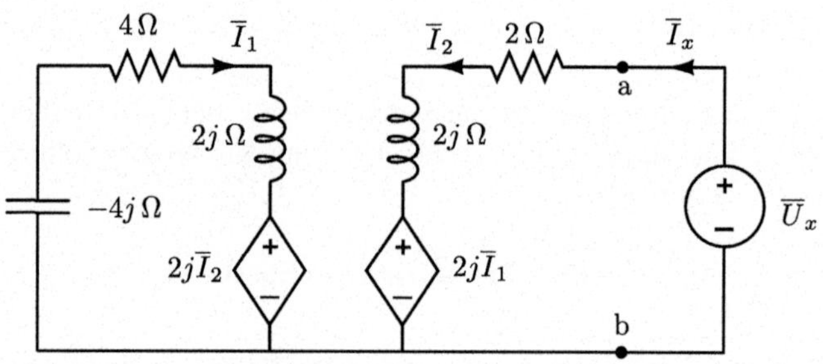

Fig. 2.212 Circuit with a test voltage source

According to Fig. 2.212, applying Kirchhoff's voltage law in the right part of the circuit yields:

$$\overline{U}_x = (2+2j)\cdot\overline{I}_2 + 2j\cdot\overline{I}_1$$

Applying Kirchhoff's voltage law again in the left part of the circuit results in:

$$(-4j+4+2j)\cdot\overline{I}_1 + 2j\overline{I}_2 = 0 \ \Rightarrow \ \overline{I}_1 = \frac{-2j}{4-2j}\cdot\overline{I}_2$$

On the other hand, $\overline{I}_2 = \overline{I}_x$, consequently:

$$\overline{U}_x = (2+2j)\cdot\overline{I}_2 + 2j\cdot\overline{I}_1 = (2+2j)\cdot\overline{I}_x + 2j\cdot\frac{-2j}{4-2j}\cdot\overline{I}_x = \left[2+2j+\frac{4}{4-2j}\right]\cdot\overline{I}_x$$

Therefore, the Thévenin impedance is as follows:

$$\overline{Z}_{th} = \frac{\overline{U}_x}{\overline{I}_x} = 2 + 2j + \frac{4}{4-2j} = \underbrace{2{,}8}_{R_{th}} + 2{,}4j \approx 3{,}69\angle 40{,}6^\circ\ \Omega$$

The value of the impedance $\overline{Z}$ that, once connected between terminals a and b, absorbs the maximum power is as follows:

$$\overline{Z} = \overline{Z}^*_{th} = 2{,}8 - 2{,}4j\ \Omega$$

Moreover, the power absorbed by said impedance is as follows:

$$P = \frac{U_{th}^2}{4 \cdot R_{th}} = \frac{7{,}16^2}{4 \cdot 2{,}8} \approx 4{,}58\ \text{W}.$$

2.59 Maximum Power Transfer

In the circuit of Fig. 2.213, the resistance R is requested to be obtained, which, once connected between terminals a and b, absorbs the maximum power. Also, calculate said power.

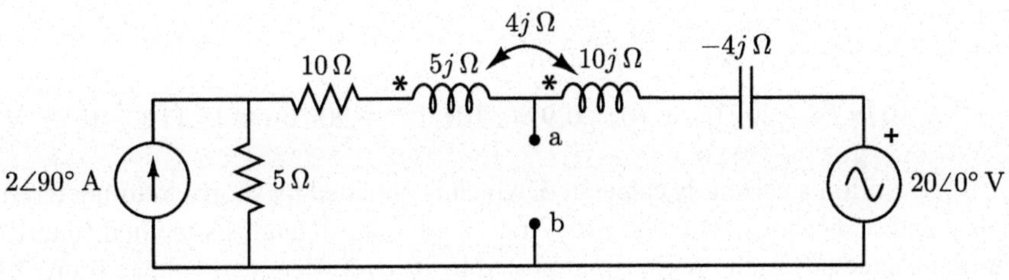

Fig. 2.213 Circuit considered in the problem under study

Solution First, the coupled coils will be replaced by their representation with dependent voltage sources in the phasor domain as shown in Fig. 2.214.

Next, the open-circuit voltage between terminals a and b, the Thévenin voltage, will be calculated. To do this, the circuit shown in Fig. 2.215 will be solved.

According to Fig. 2.215, applying Kirchhoff's voltage law (the real current source has been previously replaced by its equivalent real voltage source):

$$-10\angle 90^\circ + (5 + 10 + 5j) \cdot \overline{I}_1 + 4j\overline{I}_2 + 4j\overline{I}_1 + (10j - 4j) \cdot \overline{I}_2 + 20\angle 0^\circ = 0$$
$$\Rightarrow (15 + 9j) \cdot \overline{I}_1 + 10j \cdot \overline{I}_2 = 10\angle 90^\circ - 20\angle 0^\circ$$

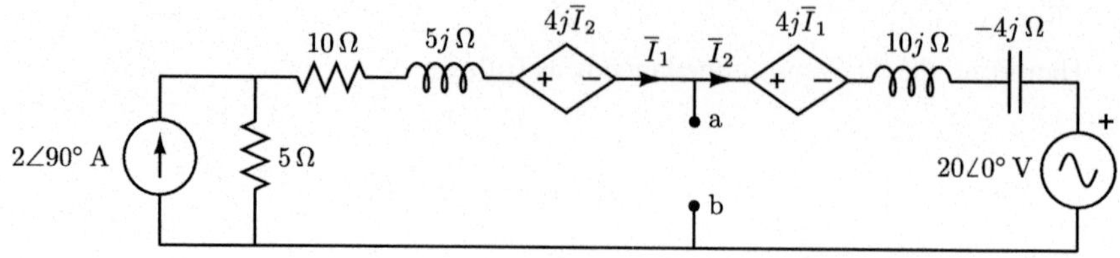

Fig. 2.214 Equivalent model of coupled coils

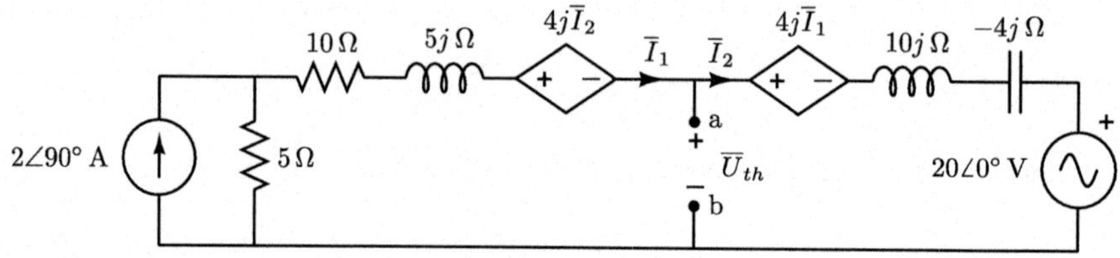

Fig. 2.215 Circuit to be solved to obtain Thévenin voltage

Taking into account that $\overline{I}_1 = \overline{I}_2$, then:

$$(15 + 9j) \cdot \overline{I}_2 + 10j \cdot \overline{I}_2 = 10\angle 90° - 20\angle 0°$$
$$\Rightarrow \overline{I}_2 = \frac{10\angle 90° - 20\angle 0°}{15 + 19j} \approx 0{,}924\angle 101{,}73° \text{ A}$$

Thus, the Thévenin voltage is as follows:

$$\begin{aligned}\overline{U}_{th} &= 4j \cdot \overline{I}_1 + (10j - 4j) \cdot \overline{I}_2 + 20\angle 0° \\ &= 10j \cdot \overline{I}_2 + 20\angle 0° = 10j \cdot 0{,}924\angle 101{,}73° + 20\angle 0° \approx 11{,}11\angle -9{,}73° \text{ V}\end{aligned}$$

Next, Norton's current is calculated, which is obtained by short-circuiting terminals a and b, as shown in Fig. 2.216. Applying the mesh analysis method (the real current source has been previously replaced by its equivalent real voltage source):

$$\begin{pmatrix} 5 + 10 + 5j & 0 \\ 0 & 10j - 4j \end{pmatrix} \begin{pmatrix} \overline{I}_a \\ \overline{I}_b \end{pmatrix} = \begin{pmatrix} 10\angle 90° - 4j\overline{I}_2 \\ -20\angle 0° - 4j\overline{I}_1 \end{pmatrix}$$

On the other hand:

$$\overline{I}_1 = \overline{I}_a \;\; ; \;\; \overline{I}_2 = \overline{I}_b$$

Substituting values

$$\begin{pmatrix} 5 + 10 + 5j & 0 \\ 0 & 10j - 4j \end{pmatrix} \begin{pmatrix} \overline{I}_a \\ \overline{I}_b \end{pmatrix} = \begin{pmatrix} 10\angle 90° - 4j\overline{I}_b \\ -20\angle 0° - 4j\overline{I}_a \end{pmatrix}$$

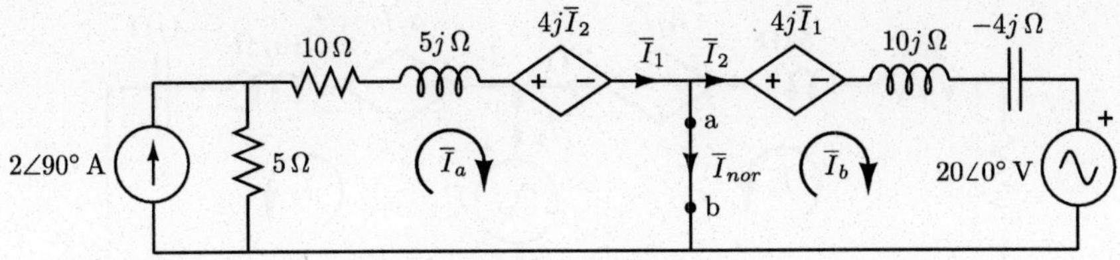

Fig. 2.216 Circuit with terminals a and b short-circuited

rearranging terms

$$\begin{pmatrix} 15+5j & 4j \\ 4j & 6j \end{pmatrix}\begin{pmatrix} \overline{I}_a \\ \overline{I}_b \end{pmatrix} = \begin{pmatrix} 10\angle 90^\circ \\ -20\angle 0^\circ \end{pmatrix}$$

and solving yields the currents $\overline{I}_a$ and $\overline{I}_b$:

$$\overline{I}_a \approx 1{,}1\angle 28{,}03^\circ\ \text{A}\ ;\ \overline{I}_b \approx 3{,}06\angle 102{,}2^\circ\ \text{A}$$

Consequently, Norton's current is as follows:

$$\overline{I}_{nor} = \overline{I}_a - \overline{I}_b = 1{,}1\angle 28{,}03^\circ - 3{,}06\angle 102{,}2^\circ \approx 2{,}96\angle -56{,}83^\circ\ \text{A}$$

Once Thévenin's voltage and Norton's current are obtained, the equivalent impedance between terminals a and b, Thévenin's impedance, is calculated as follows:

$$\overline{Z}_{th} = \frac{\overline{U}_{th}}{\overline{I}_{nor}} = \frac{11{,}11\angle -9{,}73^\circ}{2{,}96\angle -56{,}83^\circ} \approx 3{,}75\angle 47{,}1^\circ\ \Omega$$

Another way to calculate Thévenin's impedance is from the passive circuit. In this case, since the circuit has dependent sources, Thévenin's impedance cannot be obtained by impedance association. Instead, a test source connected between terminals a and b to the passive circuit as shown in Fig. 2.217 could be used. Applying the mesh analysis method:

$$\begin{pmatrix} 5+10+5j & 0 \\ 0 & 10j-4j \end{pmatrix}\begin{pmatrix} \overline{I}_a \\ \overline{I}_b \end{pmatrix} = \begin{pmatrix} -4j\overline{I}_2 - \overline{U}_x \\ -4j\overline{I}_1 + \overline{U}_x \end{pmatrix}$$

On the other hand:

$$\overline{I}_1 = \overline{I}_a\ ;\ \overline{I}_2 = \overline{I}_b$$

Substituting values

$$\begin{pmatrix} 5+10+5j & 0 \\ 0 & 10j-4j \end{pmatrix}\begin{pmatrix} \overline{I}_a \\ \overline{I}_b \end{pmatrix} = \begin{pmatrix} -4j\overline{I}_b - \overline{U}_x \\ -4j\overline{I}_a + \overline{U}_x \end{pmatrix}$$

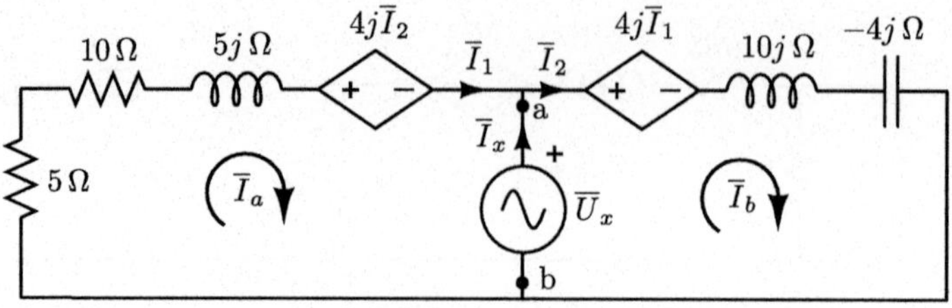

Fig. 2.217 Circuit with a test voltage source

rearranging terms

$$\begin{pmatrix} 15+5j & 4j \\ 4j & 6j \end{pmatrix} \begin{pmatrix} \overline{I}_a \\ \overline{I}_b \end{pmatrix} = \begin{pmatrix} -\overline{U}_x \\ \overline{U}_x \end{pmatrix}$$

and solving yields the currents $\overline{I}_a$ and $\overline{I}_b$:

$$\overline{I}_a \approx 0{,}11\angle 171{,}16^\circ \cdot \overline{U}_x \,\text{A} \;\; ; \;\; \overline{I}_b \approx 0{,}19\angle -67{,}88^\circ \cdot \overline{U}_x \,\text{A}$$

Considering that $\overline{I}_x = \overline{I}_b - \overline{I}_a$, then the relationship between $\overline{U}_x$ and $\overline{I}_x$ is as follows:

$$\overline{Z}_{th} = \frac{\overline{U}_x}{\overline{I}_x} = \frac{\overline{U}_x}{0{,}19\angle -67{,}88^\circ \cdot \overline{U}_x - 0{,}11\angle 171{,}16^\circ \cdot \overline{U}_x} \approx \underbrace{2{,}58}_{R_{th}} + \underbrace{2{,}77}_{X_{th}} j \,\Omega$$

Once the equivalent impedance from terminals a and b is calculated, then the resistance R that absorbs the maximum power is as follows:

$$R = \sqrt{R_{th}^2 + X_{th}^2} = \sqrt{2{,}58^2 + 2{,}77^2} \approx 3{,}78\,\Omega$$

The power absorbed by resistance R is as follows:

$$P = \frac{U_{th}^2}{2 \cdot (R + R_{th})} = \frac{11{,}11^2}{2 \cdot (3{,}78 + 2{,}58)} \approx 9{,}7\,\text{W}.$$

2.60 Maximum Power Transfer

In the circuit shown in Fig. 2.218, determine the value of the impedance $\overline{Z}$ so that, once connected between terminals a and b, it absorbs the maximum power.

Solution First, the coupled coils will be replaced by their representation with dependent voltage sources in the phasor domain as shown in Fig. 2.219.

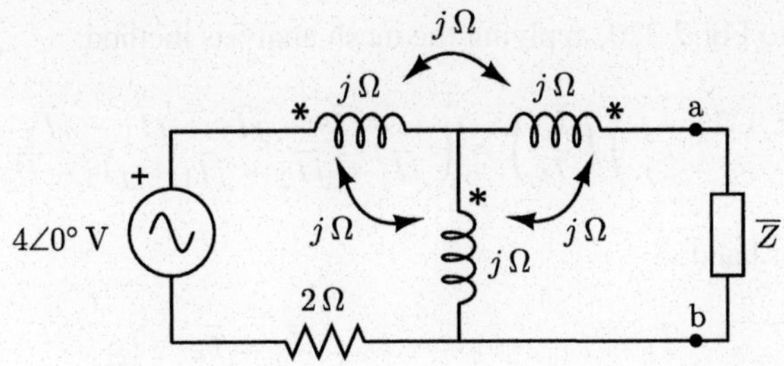

Fig. 2.218 Circuit considered in the problem under study

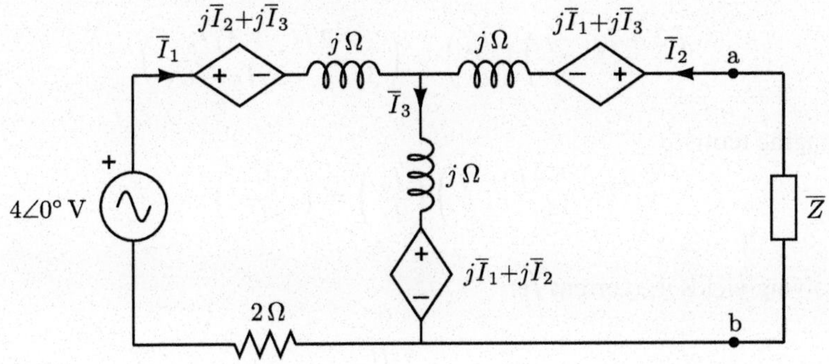

Fig. 2.219 Equivalent model of coupled coils

Next, Thévenin's impedance will be calculated. In this case, since the circuit has dependent sources, Thévenin's impedance cannot be obtained by impedance association in the passive circuit. Instead, a test source connected between terminals a and b to the circuit once the independent source is nullified, as shown in Fig. 2.220, will be used.

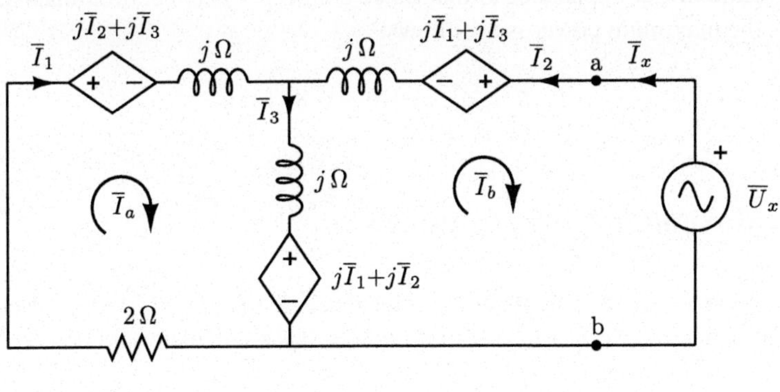

Fig. 2.220 Circuit with a test voltage source

According to Fig. 2.220, applying the mesh analysis method:

$$\begin{pmatrix} 2+2j & -j \\ -j & 2j \end{pmatrix} \begin{pmatrix} \overline{I}_a \\ \overline{I}_b \end{pmatrix} = \begin{pmatrix} -j\overline{I}_2 - j\overline{I}_3 - j\overline{I}_1 - j\overline{I}_2 \\ j\overline{I}_1 + j\overline{I}_2 + j\overline{I}_1 + j\overline{I}_3 - \overline{U}_x \end{pmatrix}$$

On the other hand:

$$\overline{I}_1 = \overline{I}_a \ ; \ \overline{I}_2 = -\overline{I}_b \ ; \ \overline{I}_3 = \overline{I}_a - \overline{I}_b$$

Substituting values

$$\begin{pmatrix} 2+2j & -j \\ -j & 2j \end{pmatrix} \begin{pmatrix} \overline{I}_a \\ \overline{I}_b \end{pmatrix} = \begin{pmatrix} -2j\overline{I}_a + 3j\overline{I}_b \\ 3j\overline{I}_a - 2j\overline{I}_b - \overline{U}_x \end{pmatrix}$$

rearranging terms

$$\begin{pmatrix} 2+4j & -4j \\ -4j & 4j \end{pmatrix} \begin{pmatrix} \overline{I}_a \\ \overline{I}_b \end{pmatrix} = \begin{pmatrix} 0 \\ -\overline{U}_x \end{pmatrix}$$

and solving yields the current $\overline{I}_b$:

$$\overline{I}_b = \frac{-\overline{U}_x}{\dfrac{16}{2+4j} + 4j}$$

Considering that $\overline{I}_b = -\overline{I}_x$, then the relationship between $\overline{U}_x$ and $\overline{I}_x$ is as follows:

$$\overline{Z}_{th} = \frac{\overline{U}_x}{\overline{I}_x} = \frac{16}{2+4j} + 4j = 1{,}6 + 0{,}8j\ \Omega$$

The value of the impedance $\overline{Z}$ that, once connected between terminals a and b, absorbs the maximum power is as follows:

$$\overline{Z} = \overline{Z}_{th}^* = 1{,}6 - 0{,}8j\ \Omega.$$

2.61 Maximum Power Transfer

In the circuit of Fig. 2.221, obtain the resistance R that, once connected between terminals a and b, absorbs the maximum power. Also, calculate said power.

Solution First, we will calculate the open-circuit voltage between terminals a and b, Thévenin voltage. To do this, we will solve the circuit shown in Fig. 2.222. According to this, applying the mesh analysis method:

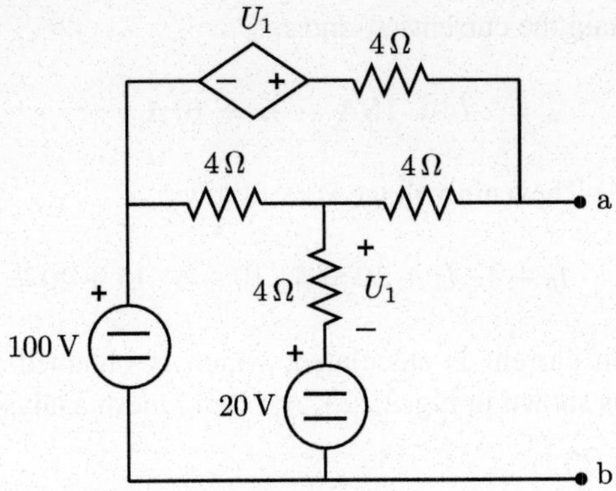

Fig. 2.221 Circuit considered in the problem under study

$$\begin{pmatrix} 8 & -4 \\ -4 & 12 \end{pmatrix} \begin{pmatrix} I_a \\ I_b \end{pmatrix} = \begin{pmatrix} 100 - 20 \\ U_1 \end{pmatrix}$$

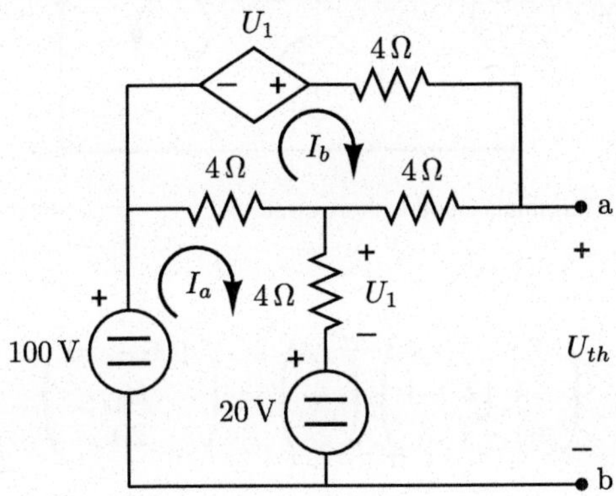

Fig. 2.222 Circuit to be solved to obtain Thévenin voltage

On the other hand:

$$U_1 = 4 \cdot I_a$$

Substituting values

$$\begin{pmatrix} 8 & -4 \\ -4 & 12 \end{pmatrix} \begin{pmatrix} I_a \\ I_b \end{pmatrix} = \begin{pmatrix} 100 - 20 \\ 4 \cdot I_a \end{pmatrix}$$

rearranging terms

$$\begin{pmatrix} 8 & -4 \\ -8 & 12 \end{pmatrix} \begin{pmatrix} I_a \\ I_b \end{pmatrix} = \begin{pmatrix} 80 \\ 0 \end{pmatrix}$$

and solving, we obtain the currents I_a and I_b:

$$I_a = 15\,\text{A} \;;\; I_b = 10\,\text{A}$$

Consequently, the Thévenin voltage is as follows:

$$U_{th} = 4 \cdot I_b + 4 \cdot I_a + 20 = 4 \cdot 10 + 4 \cdot 15 + 20 = 120\,\text{V}$$

Next, the Norton current is calculated, which is obtained by short-circuiting terminals a and b, as shown in Fig. 2.223. Applying mesh analysis:

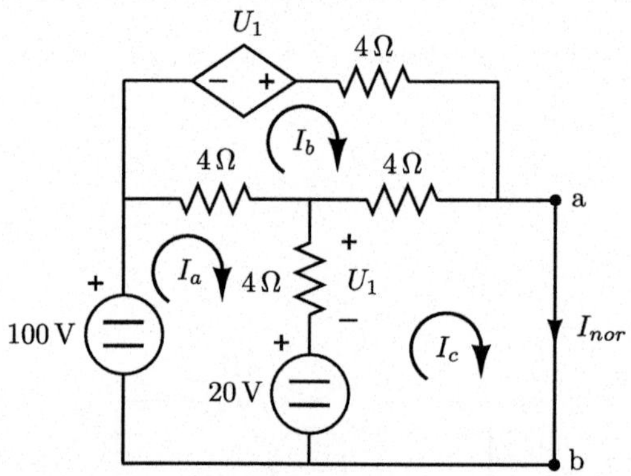

Fig. 2.223 Circuit with terminals a and b short-circuited

$$\begin{pmatrix} 8 & -4 & -4 \\ -4 & 12 & -4 \\ -4 & -4 & 8 \end{pmatrix} \begin{pmatrix} I_a \\ I_b \\ I_c \end{pmatrix} = \begin{pmatrix} 100 - 20 \\ U_1 \\ 20 \end{pmatrix}$$

On the other hand:

$$U_1 = 4 \cdot (I_a - I_c)$$

Substituting values

$$\begin{pmatrix} 8 & -4 & -4 \\ -4 & 12 & -4 \\ -4 & -4 & 8 \end{pmatrix} \begin{pmatrix} I_a \\ I_b \\ I_c \end{pmatrix} = \begin{pmatrix} 100 - 20 \\ 4 \cdot (I_a - I_c) \\ 20 \end{pmatrix}$$

rearranging terms

$$\begin{pmatrix} 8 & -4 & -4 \\ -8 & 12 & 0 \\ -4 & -4 & 8 \end{pmatrix} \begin{pmatrix} I_a \\ I_b \\ I_c \end{pmatrix} = \begin{pmatrix} 80 \\ 0 \\ 20 \end{pmatrix}$$

and solving, we obtain the current I_c, which corresponds to the Norton current, I_{nor}:

$$I_{nor} = I_c = 40\,\text{A}$$

Once the Thévenin voltage and the Norton current are obtained, the equivalent resistance between terminals a and b, Thévenin resistance, is calculated as follows:

$$R_{th} = \frac{U_{th}}{I_{nor}} = \frac{120}{40} = 3\,\Omega$$

Another way to calculate the Thévenin resistance is from the passive circuit. In this case, since the circuit has a dependent source, the Thévenin resistance cannot be obtained by associating resistors. Instead, a test source could be used connected between terminals a and b to the passive circuit, as shown in Fig. 2.224. According to the same, applying the mesh analysis method:

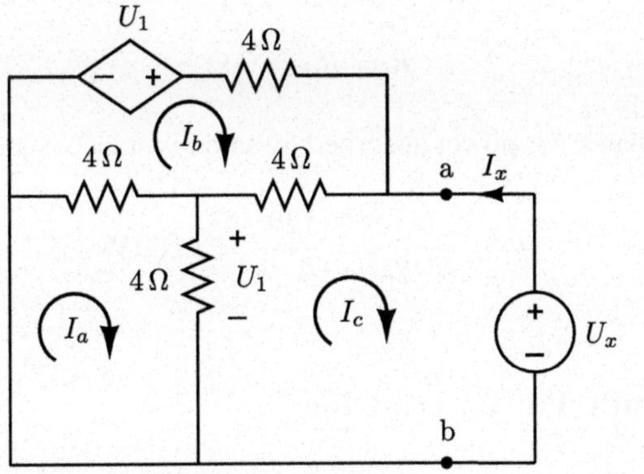

Fig. 2.224 Circuit with a test voltage source

$$\begin{pmatrix} 8 & -4 & -4 \\ -4 & 12 & -4 \\ -4 & -4 & 8 \end{pmatrix} \begin{pmatrix} I_a \\ I_b \\ I_c \end{pmatrix} = \begin{pmatrix} 0 \\ U_1 \\ -U_x \end{pmatrix}$$

On the other hand:

$$U_1 = 4 \cdot (I_a - I_c)$$

Substituting values

$$\begin{pmatrix} 8 & -4 & -4 \\ -4 & 12 & -4 \\ -4 & -4 & 8 \end{pmatrix} \begin{pmatrix} I_a \\ I_b \\ I_c \end{pmatrix} = \begin{pmatrix} 0 \\ 4 \cdot (I_a - I_c) \\ -U_x \end{pmatrix}$$

rearranging terms

$$\begin{pmatrix} 8 & -4 & -4 \\ -8 & 12 & 0 \\ -4 & -4 & 8 \end{pmatrix} \begin{pmatrix} I_a \\ I_b \\ I_c \end{pmatrix} = \begin{pmatrix} 0 \\ 0 \\ -U_x \end{pmatrix}$$

and solving, we obtain the current I_c:

$$I_c = \frac{-1}{3} \cdot U_x \text{ A}$$

Considering that $I_c = -I_x$ then the relation between U_x and I_x is as follows:

$$R_{th} = \frac{U_x}{I_x} = 3\,\Omega$$

The value of the resistance R that, once connected between terminals a and b, absorbs the maximum power is as follows:

$$R = R_{th} = 3\,\Omega$$

On the other hand, the power absorbed by said resistance is as follows:

$$P = \frac{U_{th}^2}{4 \cdot R_{th}} = \frac{120^2}{4 \cdot 3} = 1\,200\,\text{W}.$$

2.62 Maximum Power Transfer

In the circuit of Fig. 2.225, obtain the impedance $\overline{Z}$ which, once connected between terminals a and b, absorbs the maximum power. Also, calculate said power.

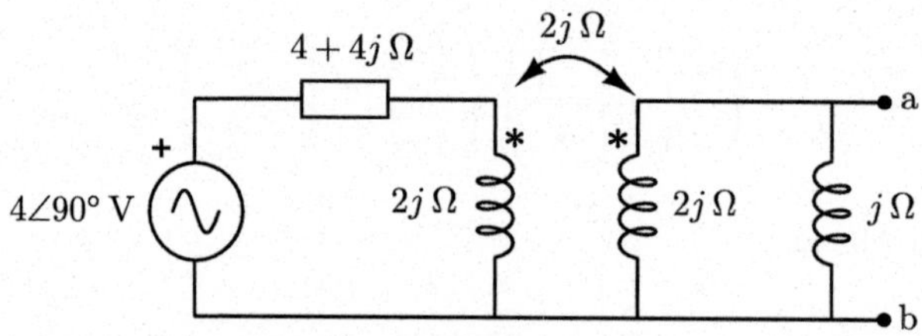

Fig. 2.225 Circuit considered in the problem under study

Solution First, the coupled coils will be replaced by their representation with dependent voltage sources in the phasor domain, as shown in Fig. 2.226.

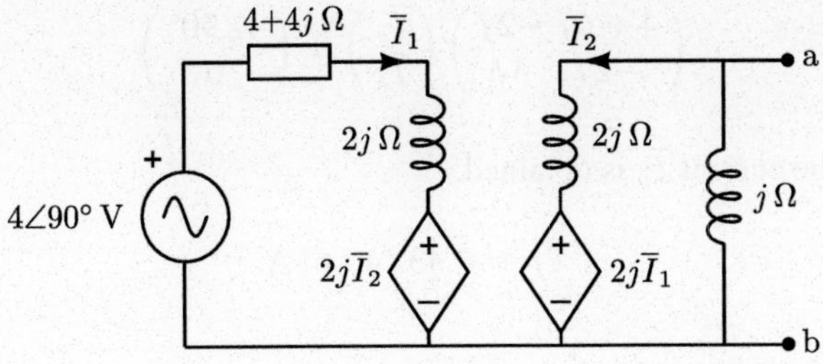

Fig. 2.226 Equivalent model of coupled coils

Next, the open-circuit voltage between terminals a and b, Thévenin voltage, will be calculated. To do this, the circuit shown in Fig. 2.227 will be solved.

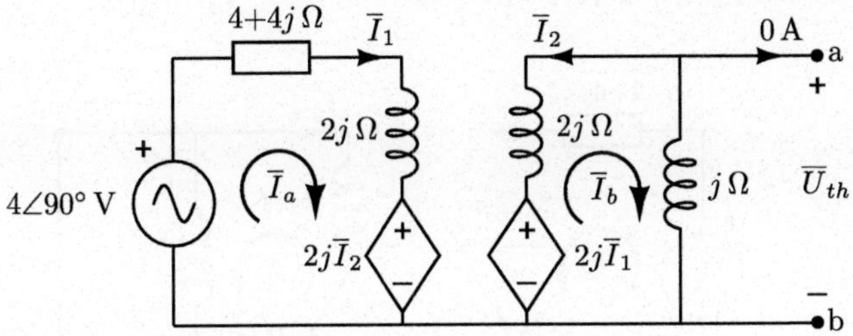

Fig. 2.227 Circuit to be solved to obtain Thévenin voltage

According to Fig. 2.227:

$$\overline{U}_{th} = -j \cdot \overline{I}_2 = j \cdot \overline{I}_b$$

Applying the mesh analysis method:

$$\begin{pmatrix} 4+6j & 0 \\ 0 & 3j \end{pmatrix} \begin{pmatrix} \overline{I}_a \\ \overline{I}_b \end{pmatrix} = \begin{pmatrix} 4\angle 90^\circ - 2j\overline{I}_2 \\ 2j\overline{I}_1 \end{pmatrix}$$

On the other hand:

$$\overline{I}_1 = \overline{I}_a \ ; \ \overline{I}_2 = -\overline{I}_b$$

Substituting values

$$\begin{pmatrix} 4+6j & 0 \\ 0 & 3j \end{pmatrix} \begin{pmatrix} \overline{I}_a \\ \overline{I}_b \end{pmatrix} = \begin{pmatrix} 4\angle 90^\circ + 2j\overline{I}_b \\ 2j\overline{I}_a \end{pmatrix}$$

rearranging terms

$$\begin{pmatrix} 4+6j & -2j \\ -2j & 3j \end{pmatrix} \begin{pmatrix} \overline{I}_a \\ \overline{I}_b \end{pmatrix} = \begin{pmatrix} 4\angle 90^\circ \\ 0 \end{pmatrix}$$

and solving, the current $\overline{I}_b$ is obtained:

$$\overline{I}_b \approx 0{,}43\angle 40{,}6^\circ \text{ A}$$

Therefore:

$$\overline{U}_{th} = j \cdot \overline{I}_b = j \cdot 0{,}43\angle 40{,}6^\circ = 0{,}43\angle 130{,}6^\circ \text{ V}$$

Next, the Norton current is calculated, which is obtained by short-circuiting terminals a and b, as shown in Fig. 2.228. It can be observed that the $j\,\Omega$ coil is short-circuited, so it will not be involved in the resolution.

According to Fig. 2.228, applying mesh analysis:

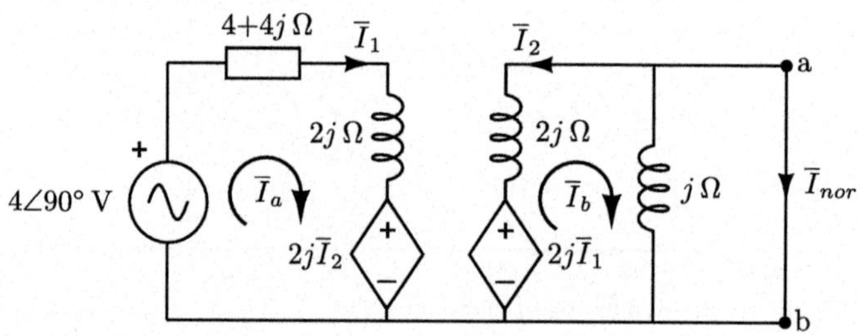

Fig. 2.228 Circuit with terminals a and b short-circuited

$$\begin{pmatrix} 4+6j & 0 \\ 0 & 2j \end{pmatrix} \begin{pmatrix} \overline{I}_a \\ \overline{I}_b \end{pmatrix} = \begin{pmatrix} 4\angle 90^\circ - 2j\overline{I}_2 \\ 2j\overline{I}_1 \end{pmatrix}$$

On the other hand:

$$\overline{I}_1 = \overline{I}_a \ ; \ \overline{I}_2 = -\overline{I}_b$$

Substituting values

$$\begin{pmatrix} 4+6j & 0 \\ 0 & 2j \end{pmatrix} \begin{pmatrix} \overline{I}_a \\ \overline{I}_b \end{pmatrix} = \begin{pmatrix} 4\angle 90^\circ + 2j\overline{I}_b \\ 2j\overline{I}_a \end{pmatrix}$$

rearranging terms

$$\begin{pmatrix} 4+6j & -2j \\ -2j & 2j \end{pmatrix} \begin{pmatrix} \overline{I}_a \\ \overline{I}_b \end{pmatrix} = \begin{pmatrix} 4\angle 90^\circ \\ 0 \end{pmatrix}$$

and solving, the current $\overline{I}_b$, which corresponds to the Norton current, $\overline{I}_{nor}$, is obtained:

$$\overline{I}_{nor} = \overline{I}_b = \frac{\sqrt{2}}{2} \angle 45^\circ \text{ A}$$

Once the Thévenin voltage and Norton current are obtained, the equivalent impedance between terminals a and b, Thévenin impedance, is calculated as follows:

$$\overline{Z}_{th} = \frac{\overline{U}_{th}}{\overline{I}_{nor}} = \frac{0{,}43\angle 130{,}6^\circ}{\frac{\sqrt{2}}{2}\angle 45^\circ} \approx 0{,}61\angle 85{,}6^\circ \approx 0{,}047 + 0{,}61j\ \Omega$$

Another way to calculate the Thévenin impedance is from the passive circuit. In this case, since the circuit has dependent sources, the Thévenin impedance cannot be obtained by impedance association. Instead, a test source could be used connected between terminals a and b to the passive circuit, as shown in Fig. 2.229.

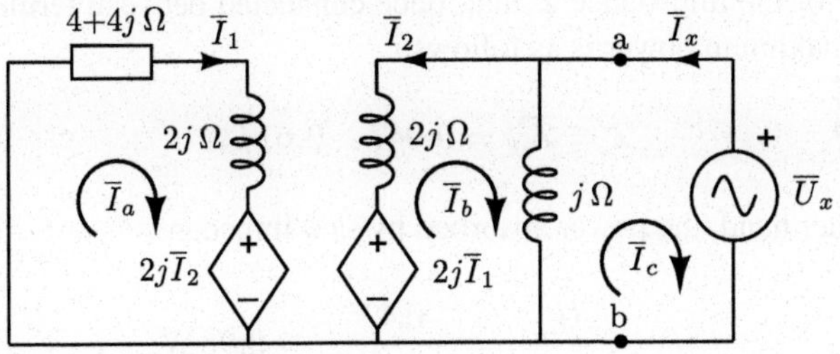

Fig. 2.229 Circuit with a test voltage source

According to Fig. 2.229, applying mesh analysis:

$$\begin{pmatrix} 4+6j & 0 & 0 \\ 0 & 3j & -j \\ 0 & -j & j \end{pmatrix} \begin{pmatrix} \overline{I}_a \\ \overline{I}_b \\ \overline{I}_c \end{pmatrix} = \begin{pmatrix} -2j\overline{I}_2 \\ 2j\overline{I}_1 \\ -\overline{U}_x \end{pmatrix}$$

On the other hand:

$$\overline{I}_1 = \overline{I}_a \ ; \ \overline{I}_2 = -\overline{I}_b$$

Substituting values

$$\begin{pmatrix} 4+6j & 0 & 0 \\ 0 & 3j & -j \\ 0 & -j & j \end{pmatrix} \begin{pmatrix} \overline{I}_a \\ \overline{I}_b \\ \overline{I}_c \end{pmatrix} = \begin{pmatrix} 2j\overline{I}_b \\ 2j\overline{I}_a \\ -\overline{U}_x \end{pmatrix}$$

rearranging terms

$$\begin{pmatrix} 4+6j & -2j & 0 \\ -2j & 3j & -j \\ 0 & -j & j \end{pmatrix} \begin{pmatrix} \overline{I}_a \\ \overline{I}_b \\ \overline{I}_c \end{pmatrix} = \begin{pmatrix} 0 \\ 0 \\ -\overline{U}_x \end{pmatrix}$$

and solving, the current $\overline{I}_c$ is obtained:

$$\overline{I}_c \approx \frac{\overline{U}_x}{-0{,}047 - 0{,}61j}$$

Taking into account that $\overline{I}_c = -\overline{I}_x$, then the relationship between $\overline{U}_x$ and $\overline{I}_x$ is as follows:

$$\overline{Z}_{th} = \frac{\overline{U}_x}{\overline{I}_x} = \underbrace{0{,}047}_{R_{th}} + 0{,}61j\ \Omega$$

The value of the impedance $\overline{Z}$ that, once connected between terminals a and b, absorbs the maximum power is as follows:

$$\overline{Z} = \overline{Z}_{th}^* = 0{,}047 - 0{,}61j\ \Omega$$

On the other hand, the power absorbed by said impedance is as follows:

$$P = \frac{U_{th}^2}{4 \cdot R_{th}} = \frac{0{,}43^2}{4 \cdot 0{,}047} \approx 0{,}98\,\mathrm{W}.$$

2.63 Maximum Power Transfer

In the circuit of Fig. 2.230, obtain the impedance $\overline{Z}$ which, once connected between terminals a and b, absorbs the maximum power. Also, calculate said power.

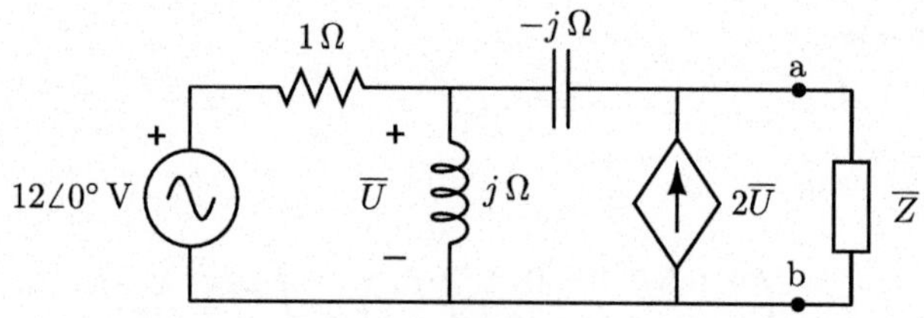

Fig. 2.230 Circuit considered in the problem under study

Solution First, the open-circuit voltage between terminals a and b, Thévenin voltage, will be calculated. To do this, the circuit shown in Fig. 2.231 will be solved.

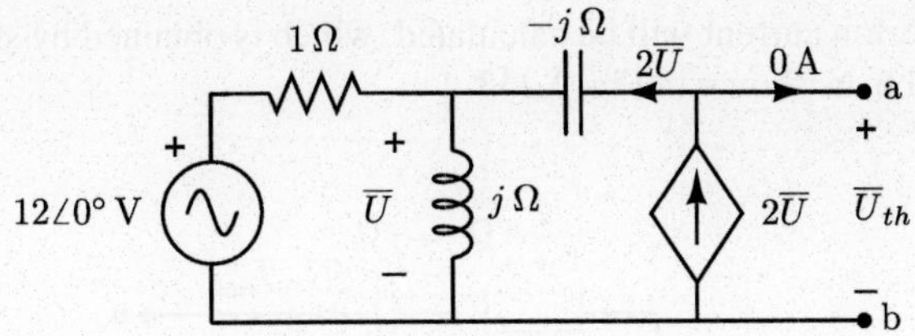

Fig. 2.231 Circuit to be solved to obtain Thévenin voltage

The circuit in Fig. 2.232 can be simplified by converting real sources, resulting in the circuit in Fig. 2.232.

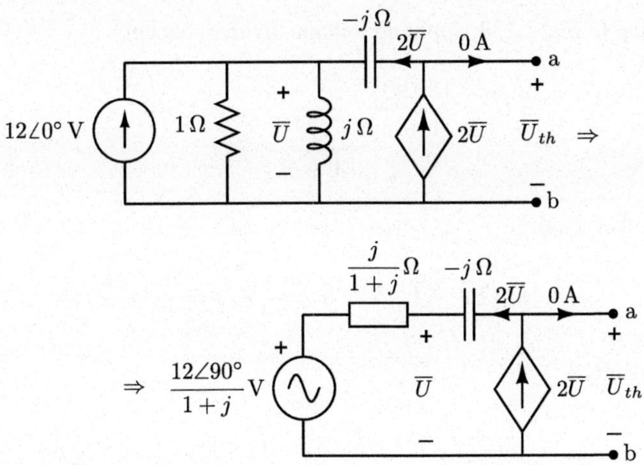

Fig. 2.232 Simplified circuit

According to Fig. 2.232:

$$\overline{U}_{th} = -j \cdot 2\overline{U} + \overline{U} = (1 - 2j) \cdot \overline{U}$$

On the other hand:

$$\overline{U} = \frac{12\angle 90°}{1+j} + \frac{j}{1+j} \cdot 2\overline{U} \;\Rightarrow\; \overline{U} = \frac{12\angle 90°}{1-j}\,\mathrm{V}$$

Consequently:

$$\overline{U}_{th} = (1 - 2j) \cdot \frac{12\angle 90°}{1-j} = 6 + 18j\,\mathrm{V}$$

Next, the Norton current will be calculated, which is obtained by short-circuiting terminals a and b, as shown in Fig. 2.233.

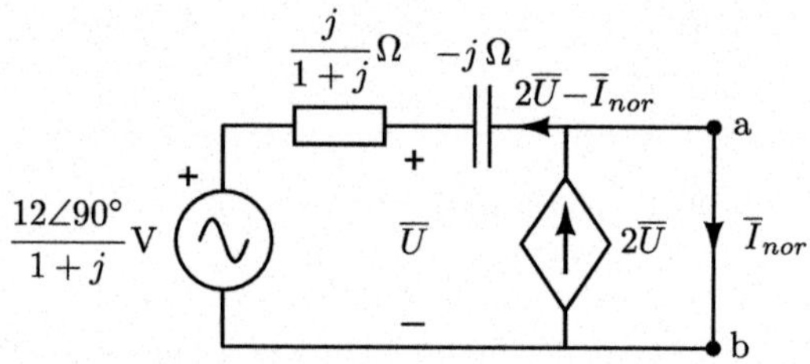

Fig. 2.233 Circuit with terminals a and b short-circuited

According to Fig. 2.233, applying voltage divider concept:

$$\overline{U} = \frac{12\angle 90°}{1+j} \cdot \frac{-j}{\dfrac{j}{1+j} - j} = 12\angle 0°$$

On the other hand:

$$\overline{U} - j \cdot (2\overline{U} - \overline{I}_{nor}) = 0 \;\Rightarrow\; \overline{I}_{nor} = \frac{1-2j}{-j} \cdot \overline{U}$$

Consequently:

$$\overline{I}_{nor} = \frac{1-2j}{-j} \cdot \overline{U} = \frac{1-2j}{-j} \cdot 12\angle 0° = 24 + 12j \text{ A}$$

Once the Thévenin voltage and Norton current are obtained, the equivalent impedance between terminals a and b, Thévenin impedance, is calculated as follows:

$$\overline{Z}_{th} = \frac{\overline{U}_{th}}{\overline{I}_{nor}} = \frac{6+18j}{24+12j} = \underbrace{0{,}5}_{R_{th}} - 0{,}5j\ \Omega$$

The value of the impedance $\overline{Z}$ that, once connected between terminals a and b, absorbs the maximum power is as follows:

$$\overline{Z} = \overline{Z}^*_{th} = 0{,}5 + 0{,}5j\ \Omega$$

On the other hand, the power absorbed by said impedance is

$$P = \frac{U^2_{th}}{4 \cdot R_{th}}$$

where

$$U_{th} = |6 + 18j| \approx 18{,}97\,\text{V}$$

Therefore:

$$P = \frac{U_{th}^2}{4 \cdot R_{th}} = \frac{18{,}97^2}{4 \cdot 0{,}5} \approx 179{,}93\,\text{W}.$$

2.64 Maximum Power Transfer

In the circuit of Fig. 2.234, obtain the impedance $\overline{Z}$ that, once connected between a and b, absorbs the maximum power. Also, calculate said power.

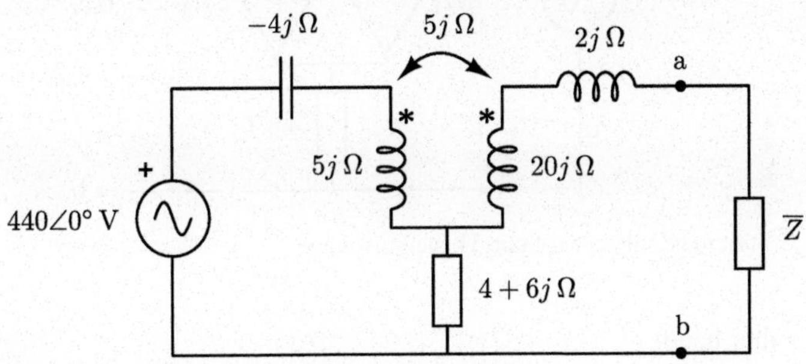

Fig. 2.234 Circuit considered in the problem under study

Solution First, the coupled coils will be replaced by their representation with phasor-dependent voltage sources, as shown in Fig. 2.235.

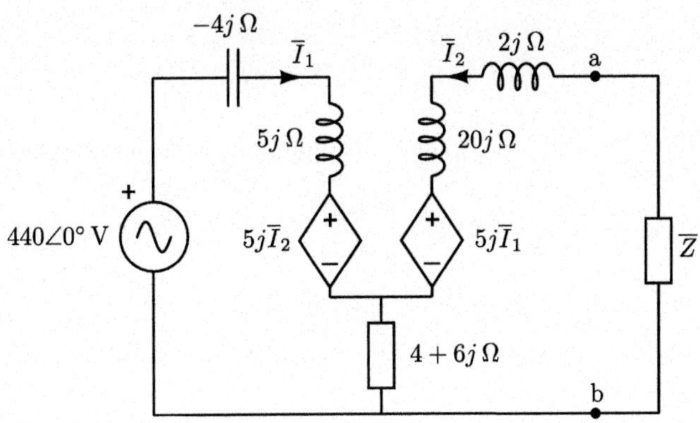

Fig. 2.235 Equivalent model of coupled coils

Next, the open-circuit voltage between terminals a and b, Thévenin voltage, will be calculated. To do this, the circuit shown in Fig. 2.236 will be solved. According to Fig. 2.236:

$$\overline{U}_{th} = (2j + 20j) \cdot \overline{I}_2 + 5j\overline{I}_1 + (4 + 6j) \cdot (\overline{I}_1 + \overline{I}_2)$$

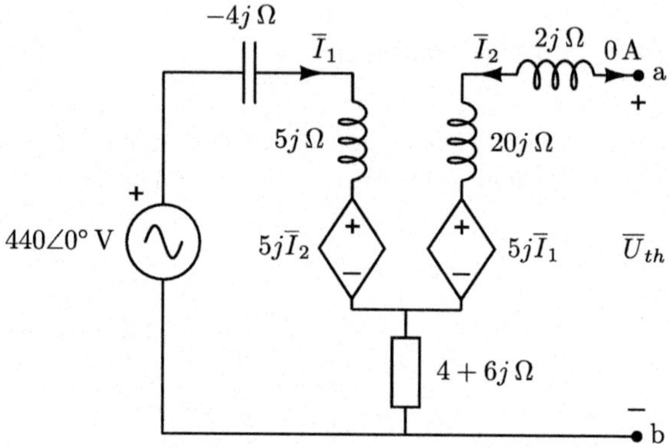

Fig. 2.236 Circuit to be solved to obtain Thévenin voltage

On the other hand:

$$440\angle 0^\circ = (5j - 4j) \cdot \overline{I}_1 + 5j\overline{I}_2 + (4 + 6j) \cdot (\overline{I}_1 + \overline{I}_2)$$

As $\overline{I}_2 = 0\,\text{A}$, then:

$$\overline{U}_{th} = (2j + 20j) \cdot \overline{I}_2 + 5j\overline{I}_1 + (4 + 6j) \cdot (\overline{I}_1 + \overline{I}_2) = (4 + 11j) \cdot \overline{I}_1 \quad (2.15)$$

and

$$440\angle 0^\circ = (5j - 4j) \cdot \overline{I}_1 + 5j\overline{I}_2 + (4 + 6j) \cdot (\overline{I}_1 + \overline{I}_2) = (4 + 7j) \cdot \overline{I}_1 \quad (2.16)$$

Solving for $\overline{I}_1$ from Eq. (2.16) and substituting its value into Eq. (2.15) yields:

$$\overline{U}_{th} = (4 + 11j) \cdot \frac{440\angle 0^\circ}{4 + 7j} \approx 638{,}79\angle 9{,}76^\circ\ \text{V}$$

Next, the Norton current will be calculated, which is obtained by short-circuiting terminals a and b, as shown in Fig. 2.237. According to Fig. 2.237, applying voltage divider concept:

$$\begin{pmatrix} 4 + 7j & -4 - 6j \\ -4 - 6j & 4 + 28j \end{pmatrix} \begin{pmatrix} \overline{I}_a \\ \overline{I}_b \end{pmatrix} = \begin{pmatrix} 440\angle 0^\circ - 5j\overline{I}_2 \\ 5j\overline{I}_1 \end{pmatrix}$$

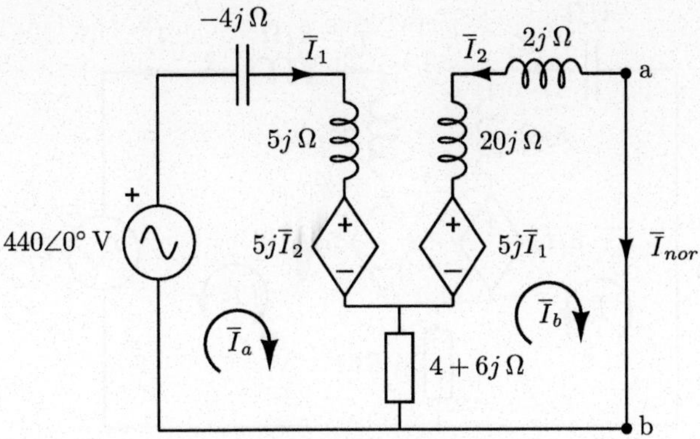

Fig. 2.237 Circuit with terminals a and b short-circuited

On the other hand:

$$\overline{I}_1 = \overline{I}_a \ ; \ \overline{I}_2 = -\overline{I}_b$$

Substituting values:

$$\begin{pmatrix} 4+7j & -4-6j \\ -4-6j & 4+28j \end{pmatrix} \begin{pmatrix} \overline{I}_a \\ \overline{I}_b \end{pmatrix} = \begin{pmatrix} 440\angle 0^\circ + 5j\overline{I}_b \\ 5j\overline{I}_a \end{pmatrix}$$

rearranging terms:

$$\begin{pmatrix} 4+7j & -4-11j \\ -4-11j & 4+28j \end{pmatrix} \begin{pmatrix} \overline{I}_a \\ \overline{I}_b \end{pmatrix} = \begin{pmatrix} 440\angle 0^\circ \\ 0 \end{pmatrix}$$

and solving yields the current $\overline{I}_b$, which corresponds to the Norton current, $\overline{I}_{nor}$:

$$\overline{I}_{nor} = \overline{I}_b = 56{,}43\angle -75{,}25^\circ \text{ A}$$

Once the Thévenin voltage and the Norton current are obtained, the equivalent impedance between terminals a and b, Thévenin impedance, is calculated as follows:

$$\overline{Z}_{th} = \frac{\overline{U}_{th}}{\overline{I}_{nor}} = \frac{638{,}79\angle 9{,}76^\circ}{56{,}43\angle -75{,}25^\circ} \approx 11{,}32\angle 85^\circ \approx 0{,}98 + 11{,}28j \ \Omega$$

Another way to calculate the Thévenin impedance is from the passive circuit. In this case, since the circuit has dependent sources, the Thévenin impedance cannot be obtained by impedance association. Instead, a test source connected between terminals a and b to the passive circuit could be used as shown in Fig. 2.238.

According to Fig. 2.238, applying the mesh analysis method:

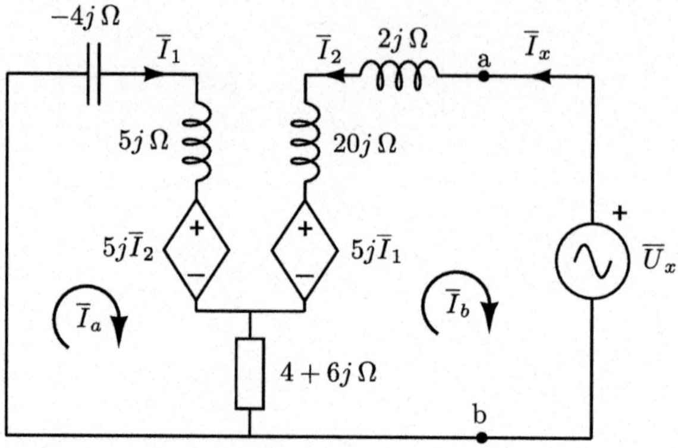

Fig. 2.238 Circuit with a test voltage source

$$\begin{pmatrix} 4+7j & -4-6j \\ -4-6j & 4+28j \end{pmatrix}\begin{pmatrix} \overline{I}_a \\ \overline{I}_b \end{pmatrix} = \begin{pmatrix} -5j\overline{I}_2 \\ 5j\overline{I}_1 - \overline{U}_x \end{pmatrix}$$

On the other hand:

$$\overline{I}_1 = \overline{I}_a \;;\; \overline{I}_2 = -\overline{I}_b$$

Substituting values:

$$\begin{pmatrix} 4+7j & -4-6j \\ -4-6j & 4+28j \end{pmatrix}\begin{pmatrix} \overline{I}_a \\ \overline{I}_b \end{pmatrix} = \begin{pmatrix} 5j\overline{I}_b \\ 5j\overline{I}_a - \overline{U}_x \end{pmatrix}$$

rearranging terms:

$$\begin{pmatrix} 4+7j & -4-11j \\ -4-11j & 4+28j \end{pmatrix}\begin{pmatrix} \overline{I}_a \\ \overline{I}_b \end{pmatrix} = \begin{pmatrix} 0 \\ -\overline{U}_x \end{pmatrix}$$

and solving yields the current $\overline{I}_b$:

$$\overline{I}_b = \frac{\overline{U}_x}{-0{,}98 - 11{,}28j}$$

Considering that $\overline{I}_b = -\overline{I}_x$, then the relationship between $\overline{U}_x$ and $\overline{I}_x$ is as follows:

$$\overline{Z}_{th} = \frac{\overline{U}_x}{\overline{I}_x} = \underbrace{0{,}98}_{R_{th}} + 11{,}28j\ \Omega$$

The value of the impedance $\overline{Z}$ that, once connected between the terminals a and b, absorbs the maximum power is as follows:

$$\overline{Z} = \overline{Z}_{th}^{*} = 0{,}98 - 11{,}28j\,\Omega$$

On the other hand, the power absorbed by this impedance is as follows:

$$P = \frac{U_{th}^2}{4 \cdot R_{th}} = \frac{638{,}79^2}{4 \cdot 0{,}98} \approx 104{,}1\,\text{kW}.$$

2.65 Maximum Power Transfer

In the circuit of Fig. 2.239, calculate the resistance R that, once connected between a and b, absorbs the maximum power. Also, obtain the value of said power.

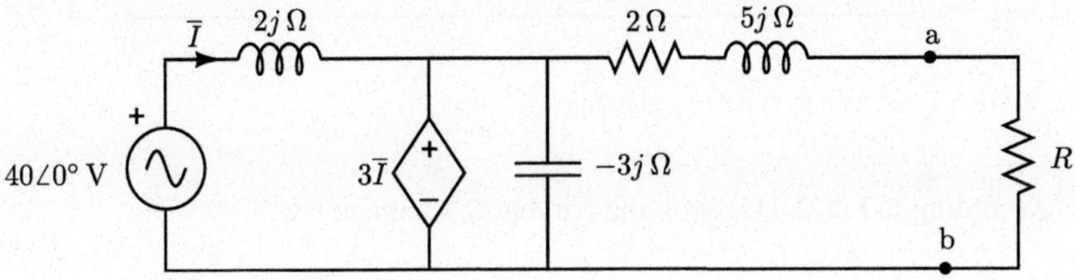

Fig. 2.239 Circuit considered in the problem under study

Solution First, the open-circuit voltage between terminals a and b, Thévenin voltage, will be obtained. To do this, the circuit shown in Fig. 2.240 will be solved. It can be observed that the resistance R has not been considered, as it is not part of the equivalent.

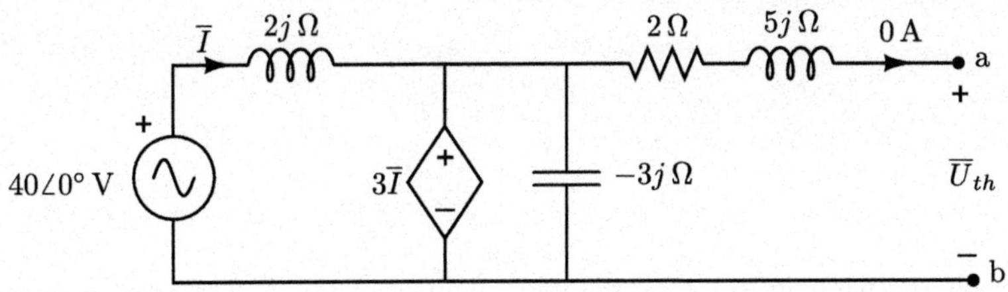

Fig. 2.240 Circuit to be solved to obtain Thévenin voltage

According to Fig. 2.240, applying Kirchhoff's voltage law:

$$40\angle 0° = 2j\overline{I} + 3\overline{I} \;\Rightarrow\; \overline{I} = \frac{40\angle 0°}{3 + 2j}\,\text{A}$$

On the other hand:

$$\overline{U}_{th} = 3\overline{I} = 3 \cdot \frac{40\angle 0^\circ}{3+2j} \approx 33{,}28\angle -33{,}69^\circ \text{ V}$$

Next, the Norton current will be calculated, which is obtained by short-circuiting terminals a and b, as shown in Fig. 2.241.

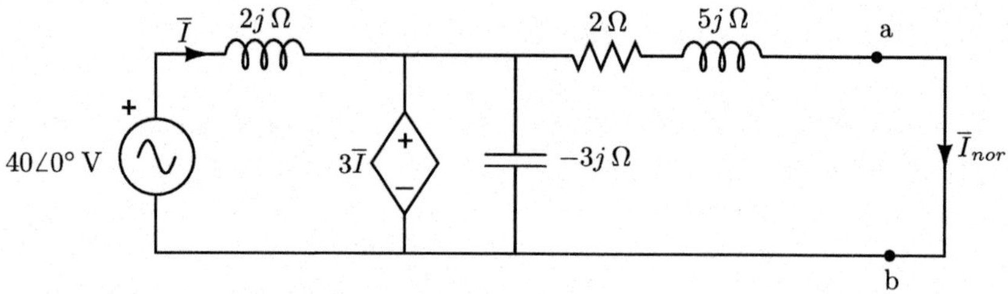

Fig. 2.241 Circuit with terminals a and b short-circuited

According to Fig. 2.241, applying Kirchhoff's voltage law:

$$40\angle 0^\circ = 2j \cdot \overline{I} + 3\overline{I} \;\Rightarrow\; \overline{I} = \frac{40\angle 0^\circ}{3+2j} \text{ A}$$

On the other hand:

$$3\overline{I} = (2+5j) \cdot \overline{I}_{nor} \;\Rightarrow\; \overline{I}_{nor} = \frac{3\overline{I}}{2+5j} = \frac{3}{2+5j} \cdot \frac{40\angle 0^\circ}{3+2j} \text{ A}$$

Once the Thévenin voltage and Norton current are obtained, the Thévenin impedance between terminals a and b is calculated as follows:

$$\overline{Z}_{th} = \frac{\overline{U}_{th}}{\overline{I}_{nor}} = \frac{3 \cdot \dfrac{40\angle 0^\circ}{3+2j}}{\dfrac{3}{2+5j} \cdot \dfrac{40\angle 0^\circ}{3+2j}} = 2+5j\,\Omega$$

An alternative way to obtain the equivalent impedance is by connecting a test source between terminals a and b, once the independent source of the circuit has been nullified, as shown in Fig. 2.242. According to this, applying Kirchhoff's voltage law:

$$2j \cdot \overline{I} + 3\overline{I} = 0 \;\Rightarrow\; \overline{I} = 0$$

On the other hand:

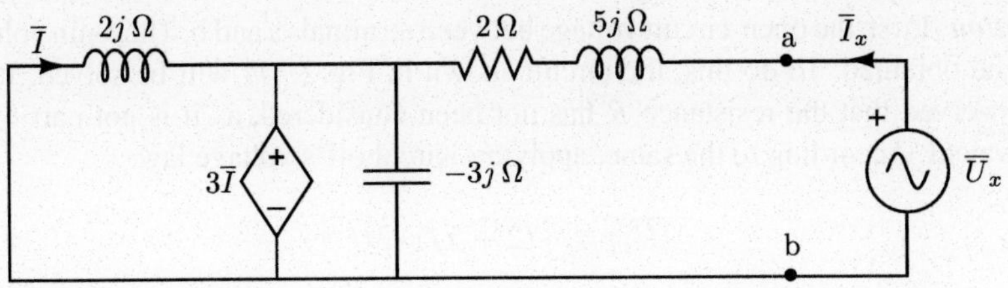

Fig. 2.242 Passive circuit

$$\overline{U}_x = (2+5j)\cdot\overline{I}_x + 3\overline{I} = (2+5j)\cdot\overline{I}_x$$

Therefore:

$$\overline{Z}_{th} = \frac{\overline{U}_x}{\overline{I}_x} = \underbrace{2}_{R_{th}} + \underbrace{5}_{X_{th}} j\,\Omega$$

The resistance R that, once connected between a and b, absorbs the maximum power, is calculated as follows:

$$R = \sqrt{R_{th}^2 + X_{th}^2} = \sqrt{2^2+5^2} = \sqrt{29}\,\Omega$$

Finally, the power absorbed by said resistance is:

$$P = \frac{U_{th}^2}{2\cdot(R_{th}+R)} = \frac{33{,}28^2}{2\cdot(2+\sqrt{29})} \approx 75\,\text{W}.$$

2.66 Maximum Power Transfer

In the circuit of Fig. 2.243, calculate the resistance R that, once connected between a and b, absorbs the maximum power. Also, obtain the value of said power.

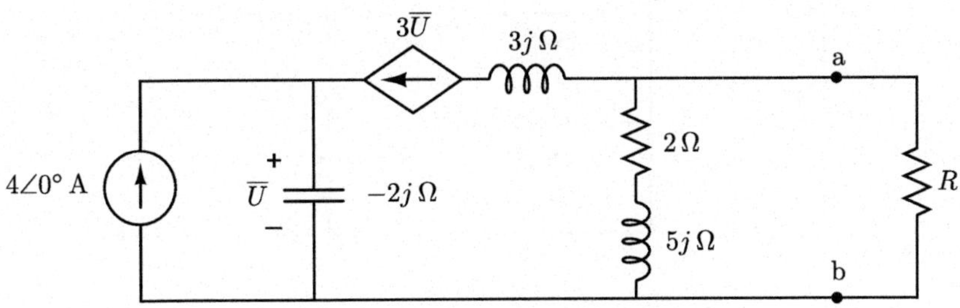

Fig. 2.243 Circuit considered in the problem under study

Solution First, the open-circuit voltage between terminals a and b, Thévenin voltage, will be obtained. To do this, the circuit shown in Fig. 2.244 will be solved. It can be observed that the resistance R has not been considered, as it is not part of the equivalent. According to the same, applying Kirchhoff's voltage law:

$$\overline{U}_{th} = -(2+5j)\cdot 3\overline{U}$$

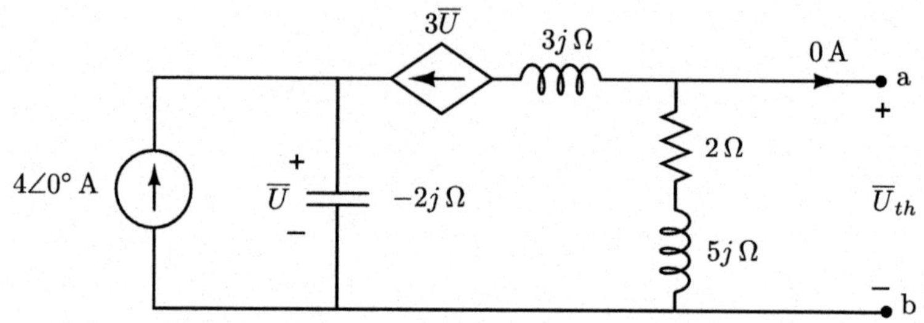

Fig. 2.244 Circuit to be solved to obtain Thévenin voltage

On the other hand:

$$\overline{U} = -2j\cdot(4\angle 0^\circ + 3\overline{U}) \;\Rightarrow\; \overline{U} = \frac{-8j}{1+6j}\ \text{V}$$

Consequently:

$$\overline{U}_{th} = -(2+5j)\cdot 3\cdot\frac{-8j}{1+6j} = 21{,}25\angle 77{,}66^\circ\ \text{V}$$

Next, the Norton current will be calculated, which is obtained by short-circuiting terminals a and b, as shown in Fig. 2.245.

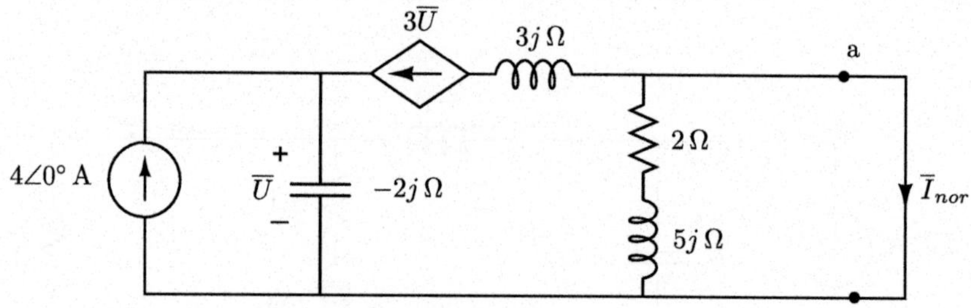

Fig. 2.245 Circuit with terminals a and b short-circuited

According to Fig. 2.245:

$$\overline{U} = -2j \cdot (4\angle 0^\circ + 3\overline{U}) \;\Rightarrow\; \overline{U} = \frac{-8j}{1+6j}\,\text{V}$$

On the other hand:

$$\overline{I}_{nor} = -3\overline{U} = -3 \cdot \frac{-8j}{1+6j}\,\text{A}$$

Once the Thévenin voltage and Norton current are obtained, the Thévenin impedance between terminals a and b is calculated as follows:

$$\overline{Z}_{th} = \frac{\overline{U}_{th}}{\overline{I}_{nor}} = \frac{-(2+5j)\cdot 3 \cdot \dfrac{-8j}{1+6j}}{-3 \cdot \dfrac{-8j}{1+6j}} = 2+5j\,\Omega$$

An alternative way to obtain the equivalent impedance is by connecting a test source between terminals a and b of the passive circuit, as shown in Fig. 2.246.

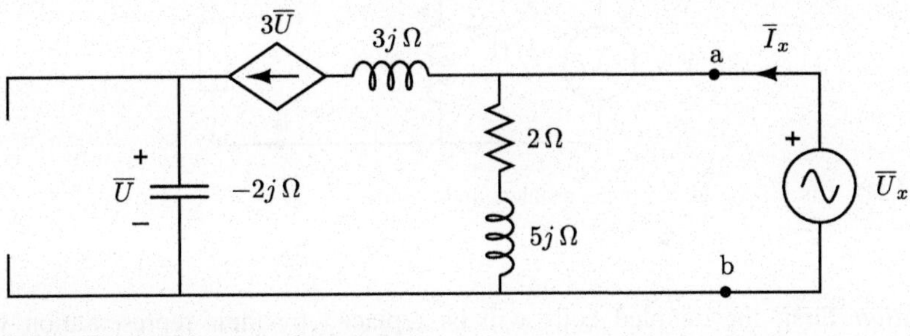

Fig. 2.246 Circuit with a test voltage source

According to Fig. 2.246:

$$-2j \cdot 3\overline{U} = \overline{U} \;\Rightarrow\; \overline{U} = 0$$

On the other hand, applying Kirchhoff's current law:

$$\overline{I}_x = \frac{\overline{U}_x}{2+5j} + 3\overline{U} = \frac{\overline{U}_x}{2+5j}$$

Therefore:

$$\overline{Z}_{th} = \frac{\overline{U}_x}{\overline{I}_x} = \underbrace{2}_{R_{th}} + \underbrace{5}_{X_{th}}\,j\,\Omega$$

The resistance R that, once connected between a and b, absorbs the maximum power is calculated as follows:

$$R = \sqrt{R_{th}^2 + X_{th}^2} = \sqrt{2^2 + 5^2} = \sqrt{29}\,\Omega$$

Finally, the power absorbed by said resistance is:

$$P = \frac{U_{th}^2}{2 \cdot (R_{th} + R)} = \frac{21{,}25^2}{2 \cdot (2 + \sqrt{29})} \approx 30{,}6\,\text{W}.$$

2.67 Maximum Power Transfer

In the circuit of Fig. 2.247, calculate the impedance $\overline{Z}$ that, once connected between terminals a and b, absorbs the maximum power. Also, obtain the value of said power.

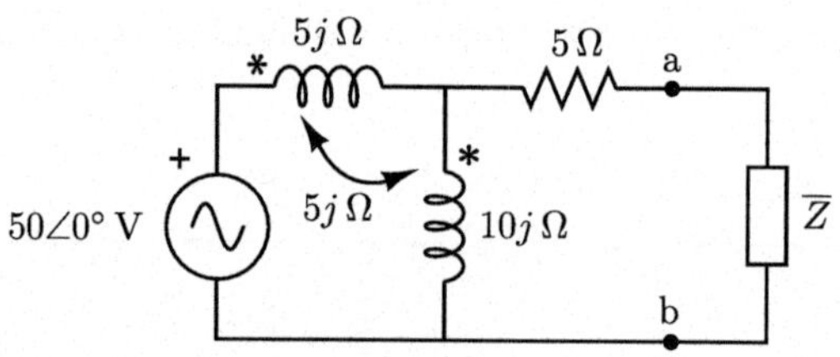

Fig. 2.247 Circuit considered in the problem under study

Solution First, the coupled coils will be replaced by their representation with dependent voltage sources in the phasor domain, as shown in Fig. 2.248.

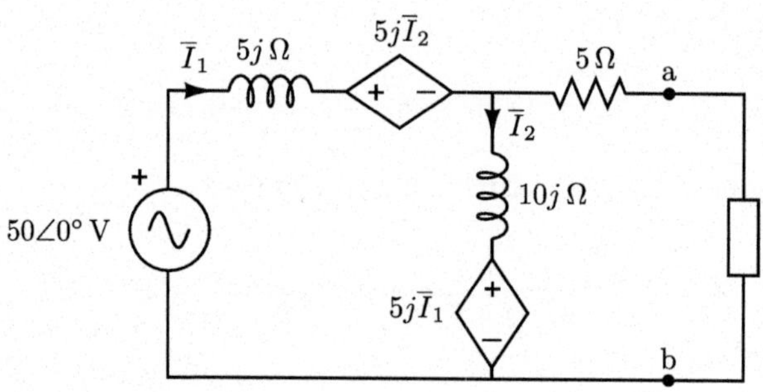

Fig. 2.248 Equivalent model of coupled coils

Next, the open-circuit voltage between terminals a and b, Thévenin voltage, will be calculated. To do this, the circuit shown in Fig. 2.249 will be solved.

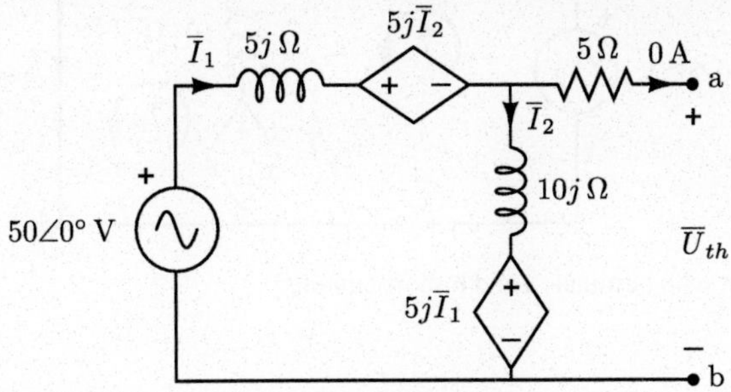

Fig. 2.249 Circuit to be solved to obtain Thévenin voltage

According to Fig. 2.249:

$$\overline{U}_{th} = 10j \cdot \overline{I}_2 + 5j \cdot \overline{I}_1$$

On the other hand:

$$50\angle 0° = 5j \cdot \overline{I}_1 + 5j\overline{I}_2 + 10j \cdot \overline{I}_2 + 5j\overline{I}_1$$

Taking into account that $\overline{I}_1 = \overline{I}_2$ then:

$$50\angle 0° = 5j \cdot \overline{I}_2 + 5j\overline{I}_2 + 10j \cdot \overline{I}_2 + 5j\overline{I}_2 \;\Rightarrow\; \overline{I}_2 = \frac{50\angle 0°}{25j} = 2\angle -90° \text{ A}$$

Thus, the Thévenin voltage is as follows:

$$\overline{U}_{th} = 10j \cdot \overline{I}_2 + 5j \cdot \overline{I}_2 = 15j \cdot \overline{I}_2 = 15j \cdot 2\angle -90° = 30\angle 0° \text{ V}$$

Next, the Norton current will be calculated, which is obtained by short-circuiting terminals a and b, as shown in Fig. 2.250.

According to Fig. 2.250, applying the mesh analysis method:

$$\begin{pmatrix} 15j & -10j \\ -10j & 5+10j \end{pmatrix} \begin{pmatrix} \overline{I}_a \\ \overline{I}_b \end{pmatrix} = \begin{pmatrix} 50\angle 0° - 5j\overline{I}_2 - 5j\overline{I}_1 \\ 5j\overline{I}_1 \end{pmatrix}$$

On the other hand:

$$\overline{I}_1 = \overline{I}_a \;;\; \overline{I}_2 = \overline{I}_a - \overline{I}_b$$

Substituting values:

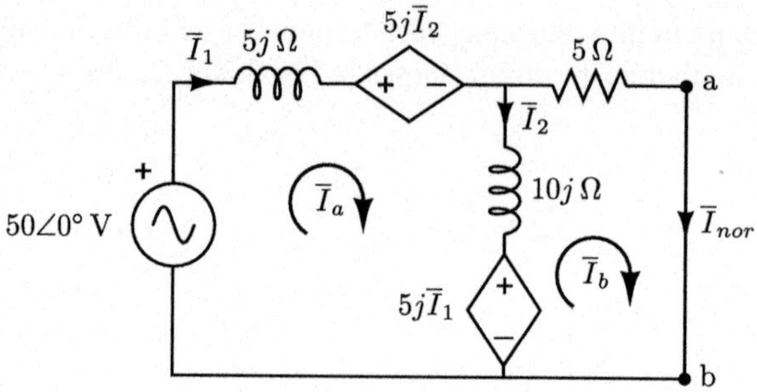

Fig. 2.250 Circuit with terminals a and b short-circuited

$$\begin{pmatrix} 15j & -10j \\ -10j & 5+10j \end{pmatrix} \begin{pmatrix} \overline{I}_a \\ \overline{I}_b \end{pmatrix} = \begin{pmatrix} 50\angle 0° - 5j \cdot (\overline{I}_a - \overline{I}_b) - 5j\overline{I}_a \\ 5j\overline{I}_a \end{pmatrix}$$

rearranging terms:

$$\begin{pmatrix} 25j & -15j \\ -15j & 5+10j \end{pmatrix} \begin{pmatrix} \overline{I}_a \\ \overline{I}_b \end{pmatrix} = \begin{pmatrix} 50\angle 0° \\ 0 \end{pmatrix}$$

and solving gives the current $\overline{I}_b$, which corresponds to the Norton current, $\overline{I}_{nor}$:

$$\overline{I}_{nor} = \overline{I}_b \approx 5{,}88\angle - 11{,}31°\text{ A}$$

Once the Thévenin voltage and Norton current are obtained, the Thévenin impedance between terminals a and b is calculated as follows:

$$\overline{Z}_{th} = \frac{\overline{U}_{th}}{\overline{I}_{nor}} = \frac{30\angle 0°}{5{,}88\angle - 11{,}31°} \approx 5{,}1\angle 11{,}31° \approx 5 + j\,\Omega$$

Another way to calculate the Thévenin impedance is from the passive circuit. In this case, since the circuit has dependent sources, the Thévenin impedance cannot be obtained by impedance association. Instead, a test source could be used connected between terminals a and b of the passive circuit, as shown in Fig. 2.251.

According to Fig. 2.251, applying the mesh analysis method:

$$\begin{pmatrix} 15j & -10j \\ -10j & 5+10j \end{pmatrix} \begin{pmatrix} \overline{I}_a \\ \overline{I}_b \end{pmatrix} = \begin{pmatrix} -5j\overline{I}_2 - 5j\overline{I}_1 \\ 5j\overline{I}_1 - \overline{U}_x \end{pmatrix}$$

On the other hand:

$$\overline{I}_1 = \overline{I}_a \;\; ; \;\; \overline{I}_2 = \overline{I}_a - \overline{I}_b$$

Substituting values:

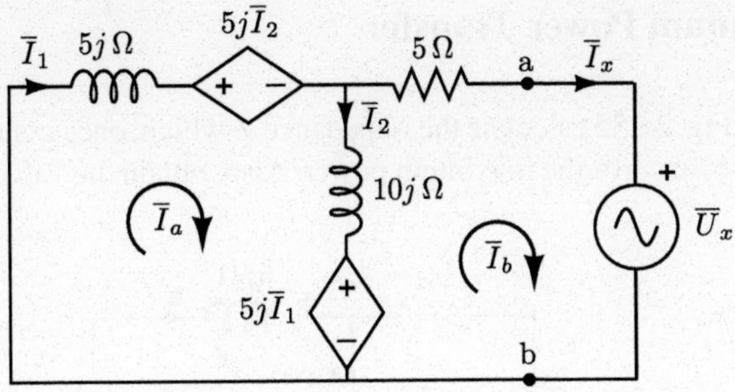

Fig. 2.251 Circuit with a test voltage source

$$\begin{pmatrix} 15j & -10j \\ -10j & 5+10j \end{pmatrix} \begin{pmatrix} \overline{I}_a \\ \overline{I}_b \end{pmatrix} = \begin{pmatrix} -5j\cdot(\overline{I}_a-\overline{I}_b)-5j\overline{I}_a \\ 5j\overline{I}_a-\overline{U}_x \end{pmatrix}$$

rearranging terms:

$$\begin{pmatrix} 25j & -15j \\ -15j & 5+10j \end{pmatrix} \begin{pmatrix} \overline{I}_a \\ \overline{I}_b \end{pmatrix} = \begin{pmatrix} 0 \\ -\overline{U}_x \end{pmatrix}$$

and solving gives the current $\overline{I}_b$:

$$\overline{I}_b \approx 0{,}1961\angle 168{,}69^\circ \cdot \overline{U}_x$$

Considering that $\overline{I}_b = -\overline{I}_x$ then the relationship between $\overline{U}_x$ and $\overline{I}_x$ is as follows:

$$\overline{Z}_{th} = \frac{\overline{U}_x}{\overline{I}_x} = \frac{-1}{0{,}1961\angle 168{,}69^\circ} \approx \underbrace{5}_{R_{th}} + j\,\Omega$$

The value of the impedance $\overline{Z}$ that, once connected between terminals a and b, absorbs the maximum power is as follows:

$$\overline{Z} = \overline{Z}_{th}^{*} = 5 - j\,\Omega$$

On the other hand, the power absorbed by this impedance will be as follows:

$$P = \frac{U_{th}^2}{4\cdot R_{th}} = \frac{30^2}{4\cdot 5} = 45\,\text{W}.$$

2.68 Maximum Power Transfer

In the circuit of Fig. 2.252 calculate the impedance $\overline{Z}$ which, once connected between terminals a and b, absorbs the maximum power. Also, obtain the value of said power.

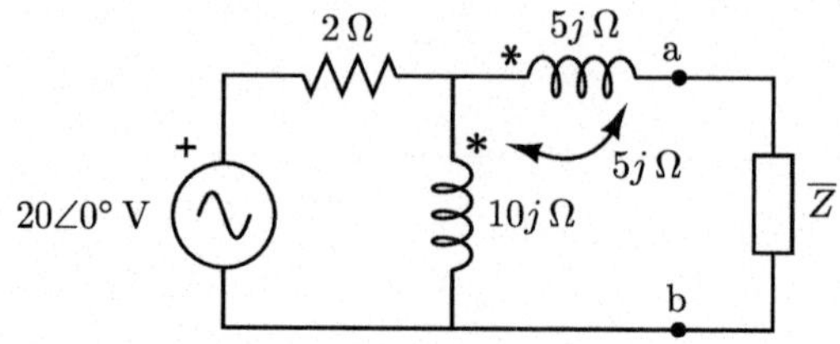

Fig. 2.252 Circuit considered in the problem under study

Solution First, the coupled coils will be replaced by their representation with dependent voltage sources in the phasor domain, as shown in Fig. 2.253. Next, the open-circuit voltage between terminals a and b, Thévenin voltage, will be calculated. To do this, the circuit shown in Fig. 2.254 will be solved, according to which:

$$\overline{U}_{th} = -5j\overline{I}_1 - 5j\overline{I}_2 + 10j \cdot \overline{I}_1 + 5j\overline{I}_2 = 5j \cdot \overline{I}_1$$

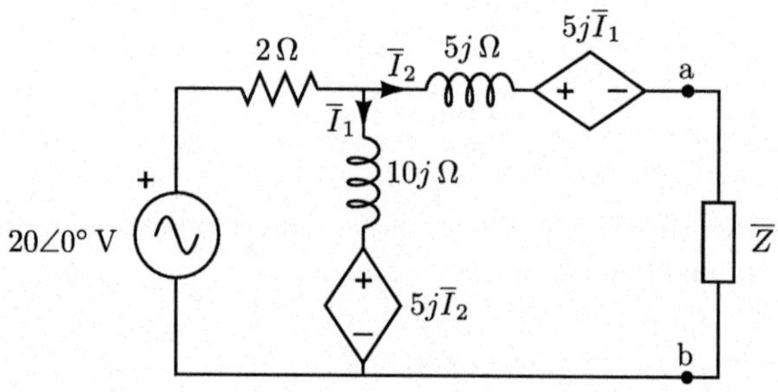

Fig. 2.253 Equivalent model of coupled coils

On the other hand:

$$20\angle 0^\circ = 2 \cdot (\overline{I}_1 + \overline{I}_2) + 10j\overline{I}_1 + 5j\overline{I}_2 = (2 + 10j) \cdot \overline{I}_1 + (2 + 5j) \cdot \overline{I}_2$$

Taking into account that $\overline{I}_2 = 0$ then:

$$20\angle 0^\circ = (2 + 10j) \cdot \overline{I}_1 \;\Rightarrow\; \overline{I}_1 = \frac{20\angle 0^\circ}{2 + 10j}$$

Therefore, the Thévenin voltage is as follows:

$$\overline{U}_{th} = 5j \cdot \overline{I}_1 = 5j \cdot \frac{20\angle 0^\circ}{2+10j} \approx 9{,}81\angle 11{,}31^\circ \text{ V}$$

Next, the Norton current will be calculated, which is obtained by short-circuiting terminals a and b, as shown in Fig. 2.255. Applying the mesh method:

$$\begin{pmatrix} 2+10j & -10j \\ -10j & 15j \end{pmatrix} \begin{pmatrix} \overline{I}_a \\ \overline{I}_b \end{pmatrix} = \begin{pmatrix} 20\angle 0^\circ - 5j\overline{I}_2 \\ 5j\overline{I}_2 - 5j\overline{I}_1 \end{pmatrix}$$

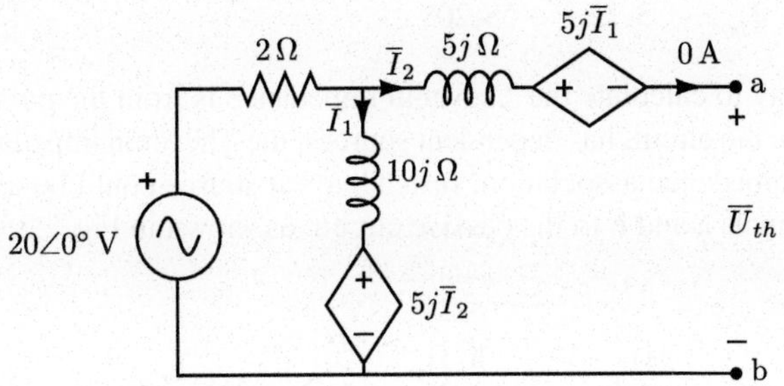

Fig. 2.254 Circuit to be solved to obtain Thévenin voltage

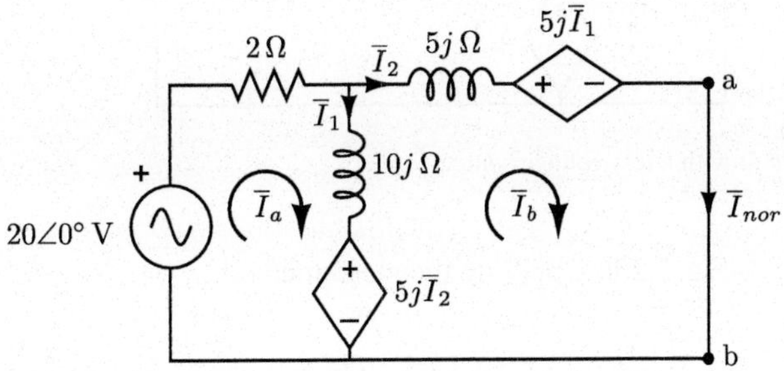

Fig. 2.255 Circuit with terminals a and b short-circuited

On the other hand:

$$\overline{I}_1 = \overline{I}_a - \overline{I}_b \ ; \ \overline{I}_2 = \overline{I}_b$$

Substituting values

$$\begin{pmatrix} 2+10j & -10j \\ -10j & 15j \end{pmatrix} \begin{pmatrix} \overline{I}_a \\ \overline{I}_b \end{pmatrix} = \begin{pmatrix} 20\angle 0^\circ - 5j\overline{I}_b \\ 5j\overline{I}_b - 5j \cdot (\overline{I}_a - \overline{I}_b) \end{pmatrix}$$

rearranging terms

$$\begin{pmatrix} 2+10j & -5j \\ -5j & 5j \end{pmatrix} \begin{pmatrix} \overline{I}_a \\ \overline{I}_b \end{pmatrix} = \begin{pmatrix} 20\angle 0^\circ \\ 0 \end{pmatrix}$$

and solving, the current $\overline{I}_b$, which corresponds to the Norton current, $\overline{I}_{nor}$, is obtained:

$$\overline{I}_{nor} = \overline{I}_b \approx 3{,}72\angle -68{,}20^\circ \text{ A}$$

Once the Thévenin voltage and the Norton current are obtained, the equivalent impedance between terminals a and b, Thévenin impedance, is calculated as follows:

$$\overline{Z}_{th} = \frac{\overline{U}_{th}}{\overline{I}_{nor}} = \frac{9{,}81\angle 11{,}31^\circ}{3{,}72\angle -68{,}20^\circ} \approx 2{,}64\angle 79{,}51^\circ \approx 0{,}48 + 2{,}59j\ \Omega$$

Another way to calculate the Thévenin impedance is from the passive circuit. In this case, since the circuit has dependent sources, the Thévenin impedance cannot be obtained by impedance association. Instead, a test source could be used connected between terminals a and b to the passive circuit, as shown in Fig. 2.256.

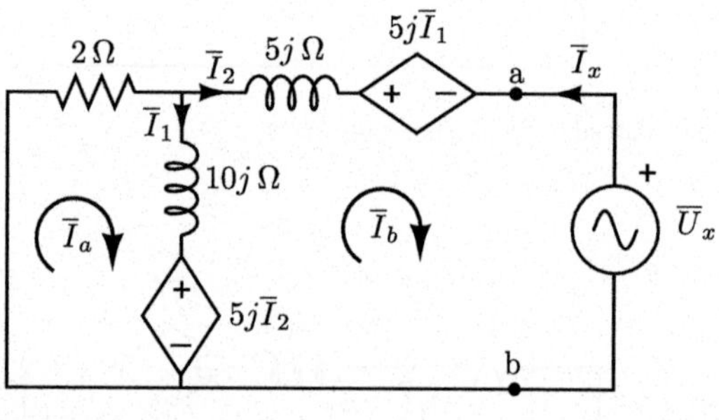

Fig. 2.256 Circuit with a test voltage source

According to Fig. 2.256, applying the mesh method:

$$\begin{pmatrix} 2+10j & -10j \\ -10j & 15j \end{pmatrix} \begin{pmatrix} \overline{I}_a \\ \overline{I}_b \end{pmatrix} = \begin{pmatrix} -5j\overline{I}_2 \\ 5j\overline{I}_2 - 5j\overline{I}_1 - \overline{U}_x \end{pmatrix}$$

On the other hand:

$$\overline{I}_1 = \overline{I}_a - \overline{I}_b \ ; \ \overline{I}_2 = \overline{I}_b$$

Substituting values

$$\begin{pmatrix} 2+10j & -10j \\ -10j & 15j \end{pmatrix} \begin{pmatrix} \overline{I}_a \\ \overline{I}_b \end{pmatrix} = \begin{pmatrix} -5j\overline{I}_b \\ 5j\overline{I}_b - 5j\cdot(\overline{I}_a - \overline{I}_b) - \overline{U}_x \end{pmatrix}$$

rearranging terms

$$\begin{pmatrix} 2+10j & -5j \\ -5j & 5j \end{pmatrix} \begin{pmatrix} \overline{I}_a \\ \overline{I}_b \end{pmatrix} = \begin{pmatrix} 0 \\ -\overline{U}_x \end{pmatrix}$$

and solving, the current $\overline{I}_b$ is obtained:

$$\overline{I}_b \approx (-0{,}069 + 0{,}372j) \cdot \overline{U}_x$$

Taking into account that $\overline{I}_b = -\overline{I}_x$ then the relationship between $\overline{U}_x$ and $\overline{I}_x$ is as follows:

$$\overline{Z}_{th} = \frac{\overline{U}_x}{\overline{I}_x} = \frac{-1}{-0{,}069 + 0{,}372j} \approx \underbrace{0{,}48}_{R_{th}} + 2{,}59j\ \Omega$$

The value of the impedance $\overline{Z}$ which, once connected between terminals a and b, absorbs the maximum power is as follows:

$$\overline{Z} = \overline{Z}_{th}^{*} = 0{,}48 - 2{,}59j\ \Omega$$

On the other hand, the power absorbed by said impedance will be as follows:

$$P = \frac{U_{th}^2}{4 \cdot R_{th}} = \frac{9{,}81^2}{4 \cdot 0{,}48} = 50{,}12\,\text{W}.$$

2.69 Maximum Power Transfer

In the circuit of Fig. 2.257, obtain the resistance R which, once connected between terminals a and b, absorbs the maximum power. Also, calculate said power.

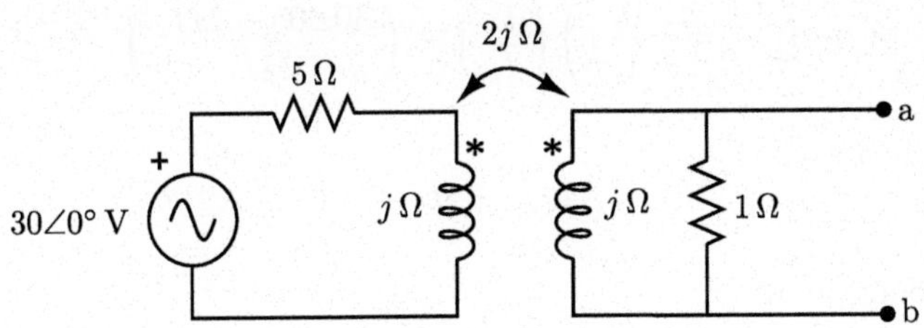

Fig. 2.257 Circuit considered in the problem under study

Solution First, the coupled coils will be replaced by their representation with dependent voltage sources in the phasor domain, as shown in Fig. 2.258.

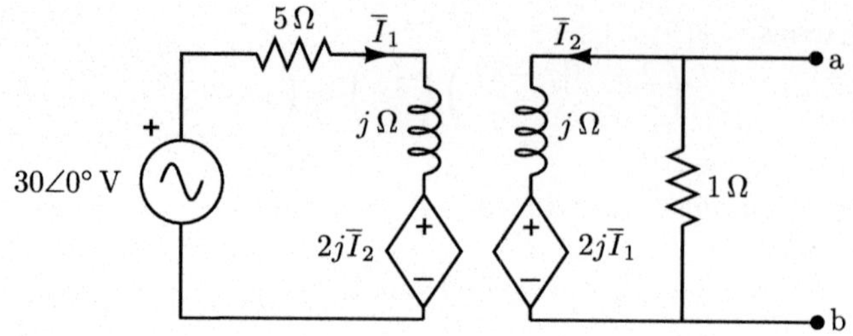

Fig. 2.258 Equivalent model of coupled coils

Next, the open-circuit voltage between terminals a and b, Thévenin voltage, will be calculated. To do this, the circuit shown in Fig. 2.259 will be solved.

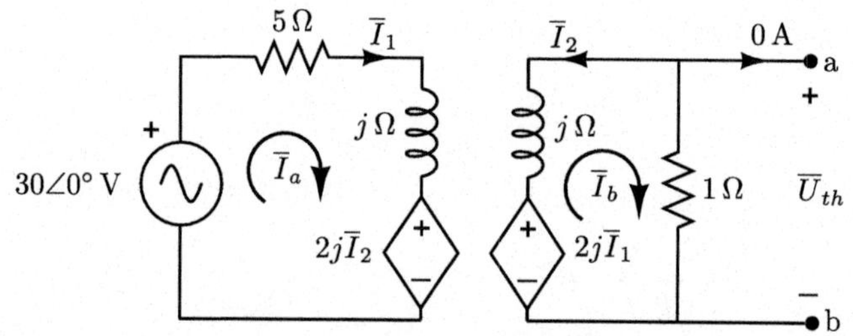

Fig. 2.259 Circuit to be solved to obtain Thévenin voltage

According to Fig. 2.259:

$$\overline{U}_{th} = -1 \cdot \overline{I}_2$$

On the other hand, applying the mesh method:

$$\begin{pmatrix} 5+j & 0 \\ 0 & 1+j \end{pmatrix} \begin{pmatrix} \overline{I}_a \\ \overline{I}_b \end{pmatrix} = \begin{pmatrix} 30\angle 0^\circ - 2j\overline{I}_2 \\ 2j\overline{I}_1 \end{pmatrix}$$

On the other hand:

$$\overline{I}_1 = \overline{I}_a \;\; ; \;\; \overline{I}_2 = -\overline{I}_b$$

Substituting values

$$\begin{pmatrix} 5+j & 0 \\ 0 & 1+j \end{pmatrix} \begin{pmatrix} \overline{I}_a \\ \overline{I}_b \end{pmatrix} = \begin{pmatrix} 30\angle 0^\circ + 2j\overline{I}_b \\ 2j\overline{I}_a \end{pmatrix}$$

rearranging terms

$$\begin{pmatrix} 5+j & -2j \\ -2j & 1+j \end{pmatrix} \begin{pmatrix} \overline{I}_a \\ \overline{I}_b \end{pmatrix} = \begin{pmatrix} 30\angle 0^\circ \\ 0 \end{pmatrix}$$

and solving, the current $\overline{I}_b$ is obtained:

$$\overline{I}_b = 3{,}6 + 4{,}8j \text{ A}$$

Since $\overline{I}_2 = -\overline{I}_b$, then:

$$\overline{U}_{th} = -1 \cdot \overline{I}_2 = -1 \cdot (-\overline{I}_b) = 3{,}6 + 4{,}8j \approx 6\angle 53{,}13° \text{ V}$$

Next, the Norton current will be calculated, which is obtained by short-circuiting terminals a and b, as shown in Fig. 2.260.

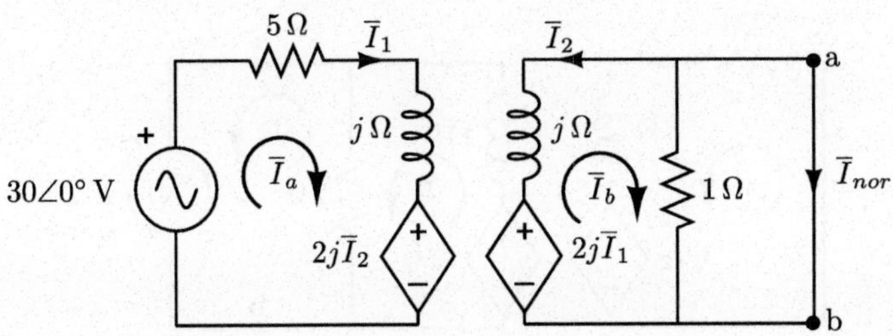

Fig. 2.260 Circuit with terminals a and b short-circuited

According to Fig. 2.260, applying the mesh method:

$$\begin{pmatrix} 5+j & 0 \\ 0 & j \end{pmatrix} \begin{pmatrix} \overline{I}_a \\ \overline{I}_b \end{pmatrix} = \begin{pmatrix} 30\angle 0° - 2j\overline{I}_2 \\ 2j\overline{I}_1 \end{pmatrix}$$

On the other hand:

$$\overline{I}_1 = \overline{I}_a \ ; \ \overline{I}_2 = -\overline{I}_b$$

Substituting values

$$\begin{pmatrix} 5+j & 0 \\ 0 & j \end{pmatrix} \begin{pmatrix} \overline{I}_a \\ \overline{I}_b \end{pmatrix} = \begin{pmatrix} 30\angle 0° + 2j\overline{I}_b \\ 2j\overline{I}_a \end{pmatrix}$$

rearranging terms

$$\begin{pmatrix} 5+j & -2j \\ -2j & j \end{pmatrix} \begin{pmatrix} \overline{I}_a \\ \overline{I}_b \end{pmatrix} = \begin{pmatrix} 30\angle 0° \\ 0 \end{pmatrix}$$

and solving, the current $\overline{I}_b$, which corresponds to the Norton current $\overline{I}_{nor}$, is obtained:

$$\overline{I}_{nor} = \overline{I}_b \approx 10{,}28\angle 30{,}96° \text{ A}$$

Once the Thévenin voltage and the Norton current are obtained, the equivalent impedance between terminals a and b, Thévenin impedance, is calculated as follows:

$$\overline{Z}_{th} = \frac{\overline{U}_{th}}{\overline{I}_{nor}} = \frac{6\angle 53{,}13^\circ}{10{,}28\angle 30{,}96^\circ} \approx 0{,}58\angle 22{,}17^\circ \approx 0{,}54 + 0{,}22j\ \Omega$$

Another way to calculate the Thévenin impedance is from the passive circuit. In this case, since the circuit has dependent sources, the Thévenin impedance cannot be obtained by impedance association. Instead, a test source could be used connected between terminals a and b to the passive circuit, as shown in Fig. 2.261.

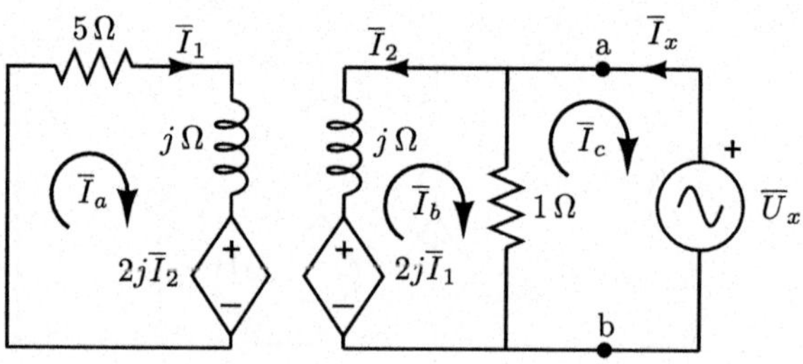

Fig. 2.261 Circuit with a test voltage source

According to Fig. 2.261, applying the mesh method:

$$\begin{pmatrix} 5+j & 0 & 0 \\ 0 & 1+j & -1 \\ 0 & -1 & 1 \end{pmatrix} \begin{pmatrix} \overline{I}_a \\ \overline{I}_b \\ \overline{I}_c \end{pmatrix} = \begin{pmatrix} -2j\overline{I}_2 \\ 2j\overline{I}_1 \\ -\overline{U}_x \end{pmatrix}$$

On the other hand:

$$\overline{I}_1 = \overline{I}_a \ ; \ \overline{I}_2 = -\overline{I}_b$$

Substituting values

$$\begin{pmatrix} 5+j & 0 & 0 \\ 0 & 1+j & -1 \\ 0 & -1 & 1 \end{pmatrix} \begin{pmatrix} \overline{I}_a \\ \overline{I}_b \\ \overline{I}_c \end{pmatrix} = \begin{pmatrix} 2j\overline{I}_b \\ 2j\overline{I}_a \\ -\overline{U}_x \end{pmatrix}$$

rearranging terms

$$\begin{pmatrix} 5+j & -2j & 0 \\ -2j & 1+j & -1 \\ 0 & -1 & 1 \end{pmatrix} \begin{pmatrix} \overline{I}_a \\ \overline{I}_b \\ \overline{I}_c \end{pmatrix} = \begin{pmatrix} 0 \\ 0 \\ -\overline{U}_x \end{pmatrix}$$

and solving, the current $\overline{I}_c$ is obtained:

$$\overline{I}_c \approx \frac{\overline{U}_x}{-0{,}54 - 0{,}22j}$$

Taking into account that $\overline{I}_c = -\overline{I}_x$, then the relationship between $\overline{U}_x$ and $\overline{I}_x$ is as follows:

$$\overline{Z}_{th} = \frac{\overline{U}_x}{\overline{I}_x} = 0{,}54 + 0{,}22j\ \Omega$$

The value of resistance R which, once connected between terminals a and b, absorbs the maximum power is as follows:

$$R = \sqrt{R_{th}^2 + X_{th}^2} = \sqrt{0{,}54^2 + 0{,}22^2} \approx 0,58\ \Omega$$

In this case, the value of the power absorbed by said resistance will be the following:

$$P = \frac{U_{th}^2}{2 \cdot (R_{th} + R)} = \frac{6^2}{2 \cdot (0{,}54 + 0{,}58)} \approx 16{,}07\ \text{W}$$

SPRINGER NATURE

GPSR Compliance

The European Union's (EU) General Product Safety Regulation (GPSR) is a set of rules that requires consumer products to be safe and our obligations to ensure this.

If you have any concerns about our products, you can contact us on ProductSafety@springernature.com

In case Publisher is established outside the EU, the EU authorized representative is:

Springer Nature Customer Service Center GmbH
Europaplatz 3
69115 Heidelberg, Germany

The manufacturer's authorised representative in the EU is Springer Nature Customer Service Centre GmbH, Europaplatz 3, 69115 Heidelberg, Germany. If you have any concerns regarding our products, please contact ProductSafety@springernature.com

Printed and bound by CPI Group (UK) Ltd, Croydon, CR0 4YY

31/07/2026

02183019-0001